CELLULAR THERAPY
CANCER
Development of Gene Therapy Based Approaches

CELLULAR THERAPY
CANCER

Development of Gene Therapy Based Approaches

Editor

Robert E Hawkins

University of Manchester, UK

World Scientific

NEW JERSEY · LONDON · SINGAPORE · BEIJING · SHANGHAI · HONG KONG · TAIPEI · CHENNAI

Published by

World Scientific Publishing Co. Pte. Ltd.
5 Toh Tuck Link, Singapore 596224
USA office: 27 Warren Street, Suite 401-402, Hackensack, NJ 07601
UK office: 57 Shelton Street, Covent Garden, London WC2H 9HE

Library of Congress Control Number: 2014936149

British Library Cataloguing-in-Publication Data
A catalogue record for this book is available from the British Library.

CELLULAR THERAPY OF CANCER
Development of Gene Therapy Based Approaches

ISBN 978-981-4295-13-0

Typeset by Stallion Press
Email: enquiries@stallionpress.com

Contents

Chapter 1   Challenges of T-Cell Therapy   1
Naomi Taylor, Anna Mondino and Balbino Alarcon

1.1   T-Cell Homeostasis   1
1.2   Lymphopenia and Immune Responsiveness
     to Self-Antigens   2
1.3   CD4 and CD8 T-Cell Differentiation States   3
1.4   T-Cell Persistence and Trafficking to Tumor   6
1.5   Choice of Target Antigen   9
References   9

Chapter 2   Gene Transfer into T Cells   19
David Gilham and Hinrich Abken

2.1   Introduction   19
2.2   γ-Retroviral Vectors to Generate
     Gene-Modified T Cells   21
2.3   The Potential Drawbacks of γ-Retroviral
     Vectors for T-Cell Gene Therapy   22
2.4   Improving All Aspects of γ-Retroviral Gene
     Transfer to Preserve or Enhance
     Antigen-Specific T-Cell Function *In Vivo*   26
2.5   Lentiviral Vectors to Transduce Minimally
     Stimulated and Quiescent Primary Human
     T Cells   29

2.6 Other Viral Vectors Used to Transduce
Primary Human T Cells   31

2.7 Non-Viral Gene Transfer Into Primary
Human T Cells: Plasmid-Based Systems   33

2.8 Non-Viral Gene Transfer Into Primary
Human T Cells: Transposon Technology   34

2.9 Non-Viral Gene Transfer Into Primary
Human T Cells: RNA Vectors   35

2.10 Mouse Models of Gene-Modified T-Cell
Therapy   38

2.11 Summary   38

References   39

Chapter 3   TCR-Engineered T cells   49
Reno Debets and Ton Schumacher

3.1 Introduction   49

3.2 Generation of TCR T Cells That are
Not or Minimally Self-Reactive   52

3.3 Generation of TCR T Cells with Enhanced
Functional Avidity   60

3.4 Next Steps in TCR Gene Therapy   67

References   70

Chapter 4   T-Bodies: Antibody-Based Engineered
T-Cell Receptors   83
John Bridgeman, Andreas A. Hombach,
David Gilham, Zelig Eshhar and Hinrich Abken

4.1 Overview   83

4.2 A Long Way to the One-Chain Format:
A Brief History of T-Bodies   84

4.3 From Structure to Function   85

4.4 Clinical Studies   103

4.5 Perspectives   111

4.6 Summary   116

References   117

Chapter 5 T-Cell Engineering and Expansion — GMP
           Issues   131
           Cor Lamers and Ryan Guest

           5.1 Introduction   131
           5.2 Clinical Vectors   133
           5.3 *Ex Vivo* Generation of Gene-Modified
                T Cells   144
           5.4 Data Management and Regulatory Aspects   163
           5.5 Future Developments   165
           References   167

Chapter 6 Clinical Trial Design   179
           Robert Hawkins, John Haanen
           and Fiona Thistlethwaite

           6.1 Introduction   179
           6.2 Background   179
           6.3 TIL Therapy   181
           6.4 Engineered T Cells — Clinical Trials   186
           6.5 Toxicities and Safety of ACT   190
           6.6 Future Strategies   193
           6.7 Conclusions   196
           References   196

Chapter 1

Challenges of T-Cell Therapy

Naomi Taylor, Anna Mondino and Balbino Alarcon

1.1 T-Cell Homeostasis

The size of the lymphocyte pool is critical for an efficient adaptive immune system; it must be sufficiently diverse to detect and destroy a wide range of potential pathogens, yet there is only limited physical space in the body to house all of these cells. Throughout life, the peripheral T-cell pool is constantly submitted to transient fluctuations in cell numbers and subset composition, yet it has long been known that regulatory mechanisms are at work to maintain a stable equilibrium. The clonal expansion of antigen-specific T cells during an immune response is subsequently followed by mass apoptosis of effector cells; only a few survive to become long-lived memory cells. Moreover, when the T-cell pool is severely depleted, the remaining T cells sense the increased availability of peripheral "space" and undergo proliferation to reconstitute steady-state T-cell numbers.[1] The overall size and composition of the peripheral T-cell pool remains fairly constant and is regulated at several levels. Thymic export seeds the peripheral tissues with newly generated T cells. Survival and proliferation through contact with self-antigens and cytokines subsequently allows their maintenance; finally, mechanisms to induce T-cell death are necessary to provide a stable equilibrium. *Homeostasis,* from the Greek words for "same" and "steady", refers to the self-regulating process that maintains the stability of a biological system; in this case, the preservation of T-cell numbers and composition over time.

In patients with tumors, the presence of a "full" lymphocyte compartment can have negative consequences for T-cell-mediated tumor immunity in that the tumor-specific T cells are not able to expand in an optimal manner. This problem is further compounded by the nature of the tumor antigen. Indeed, most identified tumor antigens represent proteins that are also expressed in non-transformed tissues and as such, mechanisms responsible for self-tolerance have ample opportunity to negatively shape the T-cell repertoire available for tumor rejection. Moreover, optimal responses by the available T cells may be further affected by tumor-induced tolerance.[2,3]

However, it appears that some of these problems can be surmounted under conditions of lymphopenia, a state in which there is a reduced number of circulating lymphocytes. In lymphopenic conditions, there is an increased homeostatic proliferation of T lymphocytes which results in the expansion of conventional T cells, in the apparent absence of antigenic stimulation.[4] Notably, this non-specific proliferation has been shown to enhance the reactivity of T cells and this effect has been exploited to augment the responsiveness of T lymphocytes in cancer patients. Indeed, the ability of adoptively-transferred autologous tumor-infiltrating lymphocytes to mount an effective response in melanoma patients has been shown to be significantly enhanced by first rendering the patients lymphopenic.[5] The mechanisms by which T cells differentiate and bypass the mechanisms of peripheral tolerance have not yet been completely clarified.

1.2 Lymphopenia and Immune Responsiveness to Self-Antigens

How does lymphopenia trigger anti-self-responses? The response is likely to be multifaceted but it is important to note that lymphopenia may result in an imbalance between effector and regulatory T cells (Tregs), with a preferential loss of the latter.[6,7] However, the loss of Tregs cannot in itself explain self-reactivity since their absence in lymphoreplete adult animals does not result in autoimmunity.[8] It is also important to note that conditioning regimens, including irradiation or chemotherapy, as well as the absence of Tregs may result in the generalized activation of antigen presenting cells (APCs), and specifically dendritic cells (DCs).[9,10]

Lymphopenia also induces the antigen-independent activation of potentially autoreactive T cells. Naïve T cells proliferate under acute lymphopenic conditions in response to the same factors that promote their survival, the interleukin-7 (IL-7) cytokine and T-cell receptor (TCR) engagement with self-peptide/major histocompatibility complex (MHC) complexes. The homeostatic proliferation of both naïve CD4 and CD8 T cells requires low affinity interactions with self-peptide/MHC complexes.[11-13] This proliferation is accompanied by a direct differentiation into memory-like T cells in the apparent absence of antigenic stimulation. Indeed, these cells are functionally and phenotypically similar to bonafide memory cells. Interestingly, it has been shown that memory-like T cells are less prone to tolerization than naïve cells, most likely due to their less stringent requirements for activation (reviewed in Refs. 14 and 15).

1.3 CD4 and CD8 T-Cell Differentiation States

Adoptive cellular immunotherapy aims to eradicate malignancies by the transfer of reactive T cells. These T cells can be derived from the tumor-bearing host and all of the following have been tested: (i) tumor-infiltrating lymphocytes, (ii) tumor-primed lymph node cells, (iii) *in vitro*-sensitized peripheral blood lymphocytes, (iv) and tumor-specific T cells generated *in vitro* by TCR/CAR (chimeric antigen receptor) gene transfer.[16-18] One important question, that had been heatedly debated until recently, revolved around the relative persistence and efficacies of naïve, central memory, and effector memory T cells for adoptive cell therapy (ACT). Central memory cells, with high proliferative and reconstituting capacity, maintain the ability to relocate to secondary lymphoid organs and are involved in recall responses. In contrast, terminally differentiated effectors, present primarily in peripheral tissues, are endowed with immediate effector function upon antigen reencounter but show poor proliferative and reconstituting abilities.[19] This is of importance as terminally differentiated effector cells might exert potent, but transient anti-tumor activity, while central memory T cells may confer a more durable T-cell immunity required for long-lasting immunosurveillance. Increasing evidence indicates that

while anti-tumor effector T cells, obtained after multiple rounds of
ex vivo stimulation, possess highly effective *in vitro* cytotoxic activity,
they are less effective than naïve or memory-like T cells *in vivo*.[20-22]

Notably, it has recently been found, in mice as well as in macaques,
that antigen-specific CD8[+] T cell derived from central memory T cells
(T_{CM}) show significantly higher long-term persistence/survival than
effector memory T cells (T_{EM}).[20,22] Moreover, these T_{CM} mediate
superior anti-tumor immunity than T_{EM} upon adoptive transfer into
tumor-bearing mice[20,22] and that stem cell memory subsets result in a
more pronounced tumor regression.[23] It is also important to note
that under conditions where a tumor-specific TCR is introduced *ex
vivo* into either naïve or central memory T cells (by retroviral-
mediated gene transfer), infusion of the former results in a significantly more robust anti-tumor response.[21] Finally, it has recently been
shown that a long-lived human memory T-cell population, characterized as CD45RO(−), CCR7(+), CD45RA(+), CD62L(+), CD27(+),
CD28(+), and IL-7Ralpha(+), has a significantly enhanced capacity
to reconstitute immunodeficient hosts and mediate an anti-tumor
response in a humanized mouse model.[24]

Considerable research efforts have thus been devoted to the
development of *ex vivo* activation strategies that preserve a naïve- or
central memory-"like" phenotype. While prolonged IL-2 signaling
promotes the terminal effector differentiation of CD8 T cells,[25,26]
homeostatic gamma chain cytokines (IL-7, IL-15, IL-21) have
proven efficacious in sustaining T-cell proliferation and *in vivo* anti-
tumor function without favoring terminal differentiation.[27,28] Of
note, it has been shown that IL-7, IL-15, and IL-21 support the
generation of CD8[+] cells with superior therapeutic activity as compared to IL-2 alone.[29-33] IL-7 and IL-15 have also been exploited for
ex vivo gene transfer; these cytokines sustain sufficient activation of
lymphocytes in the absence of TCR stimulation, rendering them
susceptible to lentiviral infection without favoring cell differentiation.[34-37] As an alternative to the use of recombinant cytokines,
genetic modification of T cells with a vector expressing a homeostatic
cytokine such as IL-15 or IL-21, may favor the generation of lymphocytes with a central memory phenotype.[38]

CD8 T cells clearly play a critical role in anti-tumor immunity, but it is important to note that there is also an important function for CD4 T cells.[39-41] Moreover, CD4 lymphocytes have recently been shown to play a direct role in adoptive immunotherapy.[42,43] Upon retroviral transfer of a tumor-specific receptor into T lymphocytes, it was demonstrated that gene-modified CD8 as well as CD4 T cells are required for an efficient immune response. The most potent anti-tumor responses were observed when the ratio of gene-modified CD4:CD8 T cells was 1:1.[44] Moreover, we and others have found that adoptively transferred memory-like CD8 T cells are subject to peripheral cross-tolerance; breakdown of this tolerance and differentiation of CD8 T cells into effectors require CD4 T cell help.[45,46] Thus, the cooperativity between CD4 and CD8 T cells needs to be taken into account in ACT protocols.

The independent role of CD4 T cells in tumor eradication was demonstrated in a model wherein irradiation alone had no effect on tumor regression but ACT, with as few as $5 \times 10e4$ tumor-specific T cells, induced an initial regression.[47,48] While this treatment did not result in a durable response in the majority of mice, it is notable that ACT of naïve tumor-specific CD4 T cells conferred a total eradication under conditions where inhibitory Tregs were inhibited by CTLA-4 blockade.[43] Furthermore, other studies have found that the initial recruitment of a Treg versus T-effector response (within the first 2–4 days) determines the outcome of the anti-tumor immune response, resulting in either tolerance or rejection.[49]

Effector T-cell subsets are distinguished, at least partially, on the basis of the cytokines they produce, resulting in distinct functions and the recruitment of diverse cell types. Naïve $CD4^+$ T cells can differentiate into various effector T subsets such as T-helper 1 (T_H1), T_H2, and T_H17 cells. T_H1 cells produce IFN-γ; T_H2 cells produce IL-4, IL-13, and IL-5; and T_H17 are characterized by the production of IL-17, IL-21, and IL-22.[50] T_H1 and T_H17 cells appear to be the most potent players in anti-tumor responsiveness as IFN-γ occupies a central role in the immune response to tumors[50] and several recent studies have demonstrated that IL-17 inhibits tumor cell growth.[50-52] In this context, it is important to note that the metabolic environment of a tumor may itself alter the differentiation state of effector T cells (Teff). Very recent work has shown that under hypoxic conditions, the upregulation of hypoxia inducible factor (HIF-1a) promotes Th17 differentiation while alternatively, suppression of HIF-1a inhibits Th17 development and promotes Treg generation.[53,54] Furthermore, cyclophosphamide administration and OX40 co-stimulatory receptor engagement mediate improved CD8[+] T-cell priming and an improved intratumoral T-effector/Treg cell ratio.[55] These types of studies, elucidating the mechanisms leading to the preferential persistence of Teff as compared to Tregs, will be critical to our efforts to enhance anti-tumor immune responses.

1.4 T-Cell Persistence and Trafficking to Tumor

Retargeting T cells to tumor antigens by the introduction of TCR/ CAR molecules and selection of those lymphocytes with regenerative capacities and long-term survival potential are likely to be important aspects of future clinical trials. However, it is important to note that this may not be sufficient to achieve a clinical response.[56] The efficacy of adoptively transferred T cells relies critically on their capacity to migrate to the sites of tumor development. T-cell migration is dependent on the expression of a combination of selectins, chemokine receptors, and integrins that regulate extravasation and trafficking in different microenvironments.[57] Thus, differential expression of adhesion and chemokine receptors on T cells (and the corresponding expression of ligands by tumor cells/stroma) will modulate their ability to provide immune surveillance and migration to the tumor. Accordingly, forced expression of defined chemokine receptors (for instance, CXCR2, CCR4) has been shown to improve migration of redirected T cells to tumor sites.[58,59]

Persistence and long-term functionality of adoptively transferred tumor-reactive T cells within the tumor-bearing host are critical for therapeutic efficacy. Modulation of the immunologic environment in the patient plays a major contribution in the engraftment, differentiation, clonal expansion, anergy, and/or death of transferred T cells. This is generally referred to as "host-conditioning", and is attained using myeloablative drugs (i.e., combinations including busulfan, cyclophosphamide, fludarabine, melphalan, and/or treosulfan) or irradiation. While it is still not completely clear how host conditioning influences the development of anti-tumor T-cell responses, the following parameters have been shown to play important roles: (1) Chemotherapy can render solid tumors accessible to transferred T cells by changing the vascular tumor endothelium,[60] and inducing tumor cell death.[61] (2) By depleting endogenous T cells, there is an increased availability of endogenous pro-survival homeostatic cytokines for the remaining endogenous T cells and/or adoptively transferred T cells, resulting in their improved functionality.[62,63] (3) Total body irradiation and/or chemotherapy damage the integrity of mucosa barriers in the intestinal tract, thereby resulting in the translocation of microbial products.[64,65] These serve as immunological adjuvants as proven by their ability to increase production of inflammatory cytokines, activate DC, and augment T-cell responses.[45,66] Thus, while the outcomes of adoptive T-cell therapies will clearly be affected by the degree of lymphopenia and regulatory T-cell depletion that are induced by conditioning regimens, this is unlikely to be the only parameter. Host stromal cell and APC populations, cytokine milieu as well as microbial flora, will also be influenced by the conditioning regimen and these factors will in turn modulate T-cell responses to tumor antigens.

Vaccination using tumor cells, tumor lysates, and tumor-antigen-derived peptides, either emulsified in adjuvant or loaded onto DCs,[67] can augment the efficacy of infused lymphocytes. Thus far, however, therapeutic vaccines have not been very successful in the clinic. This is possibly due to the clinical state of enrolled cancer patients (who have generally failed to respond to previous treatments), and to their profound state of tolerance and immune suppression.[68] However, in the context of ATT, vaccination has proven efficacious[69] and able to favor the expansion of TCR/CAR-redirected T-cell following infusion,[70,71] at least in the context of animal models. This may be due to the fact that under conditions of ATT, "fresh" lymphocytes, capable of responding to the vaccination strategy in a non-tolerizing environment, are provided. The finding that chemotherapy does not hinder the immunogenic potential of various vaccine formulations[72] and can favor an anti-tumor response[73] have prompted the implementation of combined chemo-immunotherapies.

Chemotherapy preconditioning, ACT, and vaccination followed by adjuvant peritumoral injections of immunostimulatory nucleic acids elicit potent cytotoxic inflammatory responses.[74] Chemotherapy preconditioning markedly increases the expansion and function of adoptively transferred T cells in response to vaccination while delaying the contraction phase. These effects can be further enhanced by the use of the TLR3 ligand poly(I:C), likely by favoring DC maturation.[75] In addition to chemotherapeutic drugs, local tumor irradiation combined with intratumoral DC administration can significantly enhance the therapeutic efficacy of adoptively-transferred tumor-reactive T cells.[76] Immunomodulating agents such as Toll-like receptor agonists, CD40 agonists, adjuvants targeting components of the adaptive (IL-2/anti-IL-2 mAb complexes) and innate (poly(I:C)) immune responses[77] as well as IL-7 administration[42] support the reactivation, expansion, and survival of infused cells. Similarly, the administration of antibodies that are able to neutralize immunosuppressive cytokines such as TGF-β and IL-10, inhibitors of IDO-positive DCs, and antagonistic antibodies against the inhibitory receptor CTLA-4, improve T-cell responsiveness and enhance vaccine potential.[78,79]

The promise of recombinant homeostatic cytokines has been demonstrated in animal models and more recently, clinical trials in patients have been initiated. Notably, short-term administration of recombinant IL-7 has been shown to increase the homeostatic proliferation of endogenous CD4$^+$ and CD8$^+$ lymphocytes in patients with metastatic disease.[80–82] IL-15 can similarly promote T-cell function *in vitro*[83] and *in vivo*[84,85] and patients are currently being recruited for a phase I clinical trial of intravenously-administered rIL-15 (NCI, NCT01021059). Moreover, encouraging recent results have been observed in malignant melanoma and renal cell carcinoma patients treated with rIL-21.[86,87] Thus, the use of recombinant IL-7, IL-15, and IL-21, either alone or in combination, may significantly enhance the persistence and function of adoptively transferred T cells.

1.5 Choice of Target Antigen

The choice of target antigen clearly bears on the efficacy of the tumor-reactive lymphocytes. In the context of infiltrating lymphocytes, the tumor antigen is obviously not "chosen" and likely represents a polyclonal population. This can have important consequences as it has recently been shown that when precursor frequencies of tumor-specific T cells are below a competitive threshold, the lymphocytes undergo increased proliferation and have enhanced anti-tumor efficacy, as demonstrated by an improved tumor eradication.[88] Notably though, when tumor-specific lymphocytes are primed at frequencies above this threshold, the functional benefit is impaired, possibly due to intraclonal competition.[88] Overcoming intraclonal competition, while still transferring sufficient numbers of lymphocytes that are able to target large tumor masses, may require that the transferred T cells target different antigens on the tumor or on tumor-associated stroma via receptors selected for defined MHC/peptide combinations.[89] Indeed, the rejection of spontaneous tumors by tumor-specific T cells is improved by harnessing responsiveness against minor histocompatibility antigens expressed on the tumor.[90]

References

1. Rocha, B., Dautigny, N., and Pereira, P. (1989). Peripheral T lymphocytes: Expansion potential and homeostatic regulation of pool sizes and CD4/CD8 ratios in vivo. *Eur J Immunol* 19, 905–911.
2. Ahmadzadeh, M., Johnson, L.A., Heemskerk, B., Wunderlich, J.R., Dudley, M.E., White, D.E., and Rosenberg, S.A. (2009). Tumor antigen-specific CD8 T cells infiltrating the tumor express high levels of PD-1 and are functionally impaired. *Blood* 114, 1537–1544.
3. Staveley-O'Carroll, K., Sotomayor, E., Montgomery, J., Borrello, I., Hwang, L., Fein, S., Pardoll, D., and Levitsky, H. (1998). Induction of antigen-specific T cell anergy: An early event in the course of tumor progression. *Proc Natl AcadSci USA* 95, 1178–1183.
4. Jameson, S.C. (2002). Maintaining the norm: T-cell homeostasis. *Nat Rev Immunol* 2, 547–556.
5. Dudley, M.E., Wunderlich, J.R., Robbins, P.F., Yang, J.C., Hwu, P., Schwartzentruber, D.J., Topalian, S.L., Sherry, R., Restifo, N.P., Hubicki, A.M.,

et al. (2002). Cancer regression and autoimmunity in patients after clonal repopulation with antitumor lymphocytes. *Science* 298, 850–854.

6. Antony, P.A., Piccirillo, C.A., Akpinarli, A., Finkelstein, S.E., Speiss, P.J., Surman, D.R., Palmer, D.C., Chan, C.C., Klebanoff, C.A., Overwijk, W.W., *et al.* (2005b). CD8+ T cell immunity against a tumor/self-antigen is augmented by CD4+ T helper cells and hindered by naturally occurring T regulatory cells. *J Immunol* 174, 2591–2601.

7. Brode, S., Raine, T., Zaccone, P., and Cooke, A. (2006). Cyclophosphamide-induced type-1 diabetes in the NOD mouse is associated with a reduction of CD4+CD25+Foxp3+ regulatory T cells. *J Immunol* 177, 6603–6612.

8. Lahl, K., Loddenkemper, C., Drouin, C., Freyer, J., Arnason, J., Eberl, G., Hamann, A., Wagner, H., Huehn, J., and Sparwasser, T. (2007). Selective depletion of Foxp3+ regulatory T cells induces a scurfy-like disease. *J Exp Med* 204, 57–63.

9. Paulos, C.M., Wrzesinski, C., Kaiser, A., Hinrichs, C.S., Chieppa, M., Cassard, L., Palmer, D.C., Boni, A., Muranski, P., Yu, Z., *et al.* (2007). Microbial translocation augments the function of adoptively transferred self/tumor-specific CD8+ T cells via TLR4 signaling. *J Clin Invest* 117, 2197–2204.

10. Salem, M.L., Al-Khami, A.A., El-Naggar, S.A., Diaz-Montero, C.M., Chen, Y., and Cole, D.J. (2010). Cyclophosphamide induces dynamic alterations in the host microenvironments resulting in a Flt3 ligand-dependent expansion of dendritic cells. *J Immunol* 184, 1737–1747.

11. Goldrath, A.W. and Bevan, M.J. (1999). Low-affinity ligands for the TCR drive proliferation of mature CD8+ T cells in lymphopenic hosts. *Immunity* 11, 183–190.

12. Seddon, B. and Zamoyska, R. (2002). TCR signals mediated by Src family kinases are essential for the survival of naive T cells. *J Immunol* 169, 2997–3005.

13. Surh, C.D. and Sprent, J. (2000). Homeostatic T cell proliferation: How far can T cells be activated to self-ligands? *J Exp Med* 192, F9–F14.

14. Datta, S. and Sarvetnick, N.E. (2008). IL-21 limits peripheral lymphocyte numbers through T cell homeostatic mechanisms. *PLoS One* 3, e3118.

15. Surh, C.D. and Sprent, J. (2008). Homeostasis of naive and memory T cells. *Immunity* 29, 848–862.

16. Berry, L.J., Moeller, M., and Darcy, P.K. (2009). Adoptive immunotherapy for cancer: The next generation of gene-engineered immune cells. *Tissue Antigens* 74, 277–289.

17. Dotti, G., Savoldo, B., and Brenner, M. (2009). Fifteen years of gene therapy based on chimeric antigen receptors: "Are we nearly there yet?" *Hum Gene Ther.*

18. Schmitt, T.M., Ragnarsson, G.B., and Greenberg, P.D. (2009). T cell receptor gene therapy for cancer. *Hum Gene Ther.*

19. Sallusto, F. and Lanzavecchia, A. (2009). Heterogeneity of CD4+ memory T cells: Functional modules for tailored immunity. *Eur J Immunol* 39, 2076–2082.

20. Berger, C., Jensen, M.C., Lansdorp, P.M., Gough, M., Elliott, C., and Riddell, S.R. (2008). Adoptive transfer of effector CD8+ T cells derived from central memory cells establishes persistent T cell memory in primates. *J Clin Invest* 118, 294–305.

21. Hinrichs, C.S., Borman, Z.A., Cassard, L., Gattinoni, L., Spolski, R., Yu, Z., Sanchez-Perez, L., Muranski, P., Kern, S.J., Logun, C., *et al.* (2009). Adoptively transferred effector cells derived from naive rather than central memory CD8+ T cells mediate superior antitumor immunity. *Proc Natl Acad Sci USA* 106, 17469–17474.

22. Klebanoff, C.A., Gattinoni, L., Torabi-Parizi, P., Kerstann, K., Cardones, A.R., Finkelstein, S.E., Palmer, D.C., Antony, P.A., Hwang, S.T., Rosenberg, S.A., *et al.* (2005). Central memory self/tumor-reactive CD8+ T cells confer superior antitumor immunity compared with effector memory T cells. *Proc Natl Acad Sci USA* 102, 9571–9576.

23. Klebanoff, C.A., Gattinoni, L., Palmer, D.C., Muranski, P., Ji, Y., Hinrichs, C.S., Borman, Z.A., Kerkar, S.P., Scott, C.D., Finkelstein, S.E., *et al.* (2011). Determinants of successful CD8+ T-cell adoptive immunotherapy for large established tumors in mice. *Clin Cancer Res AACR* 17, 5343–5352.

24. Gattinoni, L., Lugli, E., Ji, Y., Pos, Z., Paulos, C.M., Quigley, M.F., Almeida, J.R., Gostick, E., Yu, Z., Carpenito, C., *et al.* (2011). A human memory T cell subset with stem cell-like properties. *Nat Med* 17, 1290–1297.

25. Kalia, V., Sarkar, S., Subramaniam, S., Haining, W.N., Smith, K.A., and Ahmed, R. (2010). Prolonged interleukin-2Ralpha expression on virus-specific CD8(+) T cells favors terminal-effector differentiation in vivo. *Immunity* 32, 91–103.

26. Pipkin, M.E., Sacks, J.A., Cruz-Guilloty, F., Lichtenheld, M.G., Bevan, M.J., and Rao, A. (2010). Interleukin-2 and inflammation induce distinct transcriptional programs that promote the differentiation of effector cytolytic T cells. *Immunity* 32, 79–90.

27. Huarte, E., Fisher, J., Turk, M.J., Mellinger, D., Foster, C., Wolf, B., Meehan, K.R., Fadul, C.E., and Ernstoff, M.S. (2009). Ex vivo expansion of tumor specific lymphocytes with IL-15 and IL-21 for adoptive immunotherapy in melanoma. *Cancer Lett* 285, 80–88.

28. Kaneko, S., Mastaglio, S., Bondanza, A., Ponzoni, M., Sanvito, F., Aldrighetti, L., Radrizzani, M., La Seta-Catamancio, S., Provasi, E., Mondino, A., *et al.* (2009). IL-7 and IL-15 allow the generation of suicide gene-modified alloreactive self-renewing central memory human T lymphocytes. *Blood* 113, 1006–1015.

29. Caserta, S., Alessi, P., Basso, V., and Mondino, A. (2009). IL-7 is superior to IL-2 for ex vivo expansion of tumour-specific CD4(+) T cells. *Eur J Immunol* 40, 470–479.

30. Daudt, L., Maccario, R., Locatelli, F., Turin, I., Silla, L., Montini, E., Percivalle, E., Giugliani, R., Avanzini, M.A., Moretta, A., *et al.* (2008). Interleukin-15 favors the expansion of central memory CD8+ T cells in ex vivo generated, antileukemia human cytotoxic T lymphocyte lines. *J Immunother* 31, 385–393.
31. Hinrichs, C.S., Spolski, R., Paulos, C.M., Gattinoni, L., Kerstann, K.W., Palmer, D.C., Klebanoff, C.A., Rosenberg, S.A., Leonard, W.J., and Restifo, N.P. (2008). IL-2 and IL-21 confer opposing differentiation programs to CD8+ T cells for adoptive immunotherapy. *Blood* 111, 5326–5333.
32. Liu, S., Riley, J., Rosenberg, S., and Parkhurst, M. (2006). Comparison of common gamma-chain cytokines, interleukin-2, interleukin-7, and interleukin-15 for the in vitro generation of human tumor-reactive T lymphocytes for adoptive cell transfer therapy. *J Immunother* 29, 284–293.
33. Pouw, N., Treffers-Westerlaken, E., Kraan, J., Wittink, F., Ten Hagen, T., Verweij, J., and Debets, R. (2010). Combination of IL-21 and IL-15 enhances tumour-specific cytotoxicity and cytokine production of TCR-transduced primary T cells. *Cancer Immunol Immunother*.
34. Cavalieri, S., Cazzaniga, S., Geuna, M., Magnani, Z., Bordignon, C., Naldini, L., and Bonini, C. (2003). Human T lymphocytes transduced by lentiviral vectors in the absence of TCR activation maintain an intact immune competence. *Blood* 102, 497–505.
35. Circosta, P., Granziero, L., Follenzi, A., Vigna, E., Stella, S., Vallario, A., Elia, A.R., Gammaitoni, L., Vitaggio, K., Orso, F., *et al.* (2009). T cell receptor (TCR) gene transfer with lentiviral vectors allows efficient redirection of tumor specificity in naive and memory T cells without prior stimulation of endogenous TCR. *Hum Gene Ther* 20, 1576–1588.
36. Dardalhon, V., Jaleco, S., Kinet, S., Herpers, B., Steinberg, M., Ferrand, C., Froger, D., Leveau, C., Tiberghien, P., Charneau, P., *et al.* (2001). IL-7 differentially regulates cell cycle progression and HIV-1-based vector infection in neonatal and adult CD4+ T cells. *Proc Natl Acad Sci USA* 98, 9277–9282.
37. Verhoeyen, E., Dardalhon, V., Ducrey-Rundquist, O., Trono, D., Taylor, N., and Cosset, F.L. (2003). IL-7 surface-engineered lentiviral vectors promote survival and efficient gene transfer in resting primary T lymphocytes. *Blood* 101, 2167–2174.
38. Kaka, A.S., Shaffer, D.R., Hartmaier, R., Leen, A.M., Lu, A., Bear, A., Rooney, C.M., and Foster, A.E. (2009). Genetic modification of T cells with IL-21 enhances antigen presentation and generation of central memory tumor-specific cytotoxic T-lymphocytes. *J Immunother* 32, 726–736.
39. Antony, P.A., Piccirillo, C.A., Akpinarli, A., Finkelstein, S.E., Speiss, P.J., Surman, D.R., Palmer, D.C., Chan, C.C., Klebanoff, C.A., Overwijk, W.W., *et al.* (2005a). CD8+ T cell immunity against a tumor/self-antigen is augmented by CD4+ T helper cells and hindered by naturally occurring T regulatory cells. *J Immunol* 174, 2591–2601.

40. Benigni, F., Zimmermann, V.S., Hugues, S., Caserta, S., Basso, V., Rivino, L., Ingulli, E., Malherbe, L., Glaichenhaus, N., and Mondino, A. (2005). Phenotype and homing of CD4 tumor-specific T cells is modulated by tumor bulk. *J Immunol* 175, 739–748.

41. Pardoll, D.M. and Topalian, S.L. (1998). The role of CD4+ T cell responses in antitumor immunity. *Curr Opin Immunol* 10, 588–594.

42. Pellegrini, M., Calzascia, T., Elford, A.R., Shahinian, A., Lin, A.E., Dissanayake, D., Dhanji, S., Nguyen, L.T., Gronski, M.A., Morre, M., *et al.* (2009). Adjuvant IL-7 antagonizes multiple cellular and molecular inhibitory networks to enhance immunotherapies. *Nat Med* 15, 528–536.

43. Quezada, S.A., Simpson, T.R., Peggs, K.S., Merghoub, T., Vider, J., Fan, X., Blasberg, R., Yagita, H., Muranski, P., Antony, P.A., *et al.* (2010b). Tumor-reactive CD4+ T cells develop cytotoxic activity and eradicate large established melanoma after transfer into lymphopenic hosts. *J Exp Med.*

44. Moeller, M., Haynes, N.M., Kershaw, M.H., Jackson, J.T., Teng, M.W., Street, S.E., Cerutti, L., Jane, S.M., Trapani, J.A., Smyth, M.J., *et al.* (2005). Adoptive transfer of gene-engineered CD4+ helper T cells induces potent primary and secondary tumor rejection. *Blood* 106, 2995–3003.

45. Hamilton, S.E. and Jameson, S.C. (2008). The nature of the lymphopenic environment dictates protective function of homeostatic-memory CD8+ T cells. *Proc Natl Acad Sci USA* 105, 18484–18489.

46. Le Saout, C., Mennechet, S., Taylor, N., and Hernandez, J. (2008). Memory-like CD8+ and CD4+ T cells cooperate to break peripheral tolerance under lymphopenic conditions. *Proc Natl Acad Sci USA* 105, 19414–19419.

47. Quezada, S.A., Simpson, T.R., Peggs, K.S., Merghoub, T., Vider, J., Fan, X., Blasberg, R., Yagita, H., Muranski, P., Antony, P.A., *et al.* (2010a). Tumor-reactive CD4(+) T cells develop cytotoxic activity and eradicate large established melanoma after transfer into lymphopenic hosts. *J Exp Med* 207, 637–650.

48. Xie, Y., Akpinarli, A., Maris, C., Hipkiss, E.L., Lane, M., Kwon, E.K., Muranski, P., Restifo, N.P., and Antony, P.A. (2010). Naive tumor-specific CD4(+) T cells differentiated in vivo eradicate established melanoma. *J Exp Med* 207, 651–667.

49. Darrasse-Jeze, G., Bergot, A.S., Durgeau, A., Billiard, F., Salomon, B.L., Cohen, J.L., Bellier, B., Podsypanina, K., and Klatzmann, D. (2009). Tumor emergence is sensed by self-specific CD44hi memory Tregs that create a dominant tolerogenic environment for tumors in mice. *J Clin Invest* 119, 2648–2662.

50. Dunn, G.P., Koebel, C.M., and Schreiber, R.D. (2006). Interferons, immunity and cancer immunoediting. *Nat Rev Immunol* 6, 836–848.

51. Benchetrit, F., Ciree, A., Vives, V., Warnier, G., Gey, A., Sautes-Fridman, C., Fossiez, F., Haicheur, N., Fridman, W.H., and Tartour, E. (2002). Interleukin-17 inhibits tumor cell growth by means of a T-cell-dependent mechanism. *Blood* 99, 2114–2121.

52. Martin-Orozco, N., Muranski, P., Chung, Y., Yang, X.O., Yamazaki, T., Lu, S., Hwu, P., Restifo, N.P., Overwijk, W.W., and Dong, C. (2009). T helper 17 cells promote cytotoxic T cell activation in tumor immunity. *Immunity* 31, 787–798.

53. Dang, E.V., Barbi, J., Yang, H.Y., Jinasena, D., Yu, H., Zheng, Y., Bordman, Z., Fu, J., Kim, Y., Yen, H.R., *et al.* (2011). Control of T(H)17/T(reg) balance by hypoxia-inducible factor 1. *Cell* 146, 772–784.

54. Shi, L.Z., Wang, R., Huang, G., Vogel, P., Neale, G., Green, D.R., and Chi, H. (2011). HIF1{alpha}-dependent glycolytic pathway orchestrates a metabolic checkpoint for the differentiation of TH17 and Treg cells. *J Exp Med* 208, 1367–1376.

55. Hirschhorn-Cymerman, D., Rizzuto, G.A., Merghoub, T., Cohen, A.D., Avogadri, F., Lesokhin, A.M., Weinberg, A.D., Wolchok, J.D., and Houghton, A.N. (2009). OX40 engagement and chemotherapy combination provides potent antitumor immunity with concomitant regulatory T cell apoptosis. *J Exp Med* 206, 1103–1116.

56. Walker, R.E., Bechtel, C.M., Natarajan, V., Baseler, M., Hege, K.M., Metcalf, J.A., Stevens, R., Hazen, A., Blaese, R.M., Chen, C.C., *et al.* (2000). Long-term in vivo survival of receptor-modified syngeneic T cells in patients with human immunodeficiency virus infection. *Blood* 96, 467–474.

57. Butcher, E.C. and Picker, L.J. (1996). Lymphocyte homing and homeostasis. *Science* 272, 60–66.

58. Di Stasi, A., De Angelis, B., Rooney, C.M., Zhang, L., Mahendravada, A., Foster, A.E., Heslop, H.E., Brenner, M.K., Dotti, G., and Savoldo, B. (2009). T lymphocytes coexpressing CCR4 and a chimeric antigen receptor targeting CD30 have improved homing and antitumor activity in a Hodgkin tumor model. *Blood* 113, 6392–6402.

59. Kershaw, M.H., Wang, G., Westwood, J.A., Pachynski, R.K., Tiffany, H.L., Marincola, F.M., Wang, E., Young, H.A., Murphy, P.M., and Hwu, P. (2002). Redirecting migration of T cells to chemokine secreted from tumors by genetic modification with CXCR2. *Hum Gene Ther* 13, 1971–1980.

60. Ganss, R., Ryschich, E., Klar, E., Arnold, B., and Hammerling, G.J. (2002). Combination of T-cell therapy and trigger of inflammation induces remodeling of the vasculature and tumor eradication. *Cancer Res* 62, 1462–1470.

61. Nowak, A.K., Robinson, B.W., and Lake, R.A. (2003). Synergy between chemotherapy and immunotherapy in the treatment of established murine solid tumors. *Cancer Res* 63, 4490–4496.

62. Gattinoni, L., Finkelstein, S.E., Klebanoff, C.A., Antony, P.A., Palmer, D.C., Spiess, P.J., Hwang, L.N., Yu, Z., Wrzesinski, C., Heimann, D.M., *et al.* (2005). Removal of homeostatic cytokine sinks by lymphodepletion enhances the efficacy of adoptively transferred tumor-specific CD8+ T cells. *J Exp Med* 202, 907–912.

63. Muranski, P., Boni, A., Wrzesinski, C., Citrin, D.E., Rosenberg, S.A., Childs, R., and Restifo, N.P. (2006). Increased intensity lymphodepletion and adoptive immunotherapy--How far can we go? *Nat Clin Pract Oncol* 3, 668–681.

64. Abad, J.D., Wrzensinski, C., Overwijk, W., De Witte, M.A., Jorritsma, A., Hsu, C., Gattinoni, L., Cohen, C.J., Paulos, C.M., Palmer, D.C., *et al.* (2008). T-cell receptor gene therapy of established tumors in a murine melanoma model. *J Immunother* 31, 1–6.

65. Nakayama, M., Itoh, K., and Takahashi, E. (1997). Cyclophosphamide-induced bacterial translocation in Escherichia coli C25-monoassociated specific pathogen-free mice. *Microbiol Immunol* 41, 587–593.

66. Kieper, W.C., Troy, A., Burghardt, J.T., Ramsey, C., Lee, J.Y., Jiang, H.Q., Dummer, W., Shen, H., Cebra, J.J., and Surh, C.D. (2005). Recent immune status determines the source of antigens that drive homeostatic T cell expansion. *J Immunol* 174, 3158–3163.

67. Pilla, L., Rivoltini, L., Patuzzo, R., Marrari, A., Valdagni, R., and Parmiani, G. (2009). Multipeptide vaccination in cancer patients. *Expert Opin Biol Ther* 9, 1043–1055.

68. Rosenberg, S.A. (2001). Progress in human tumour immunology and immunotherapy. *Nature* 411, 380–384.

69. Koike, N., Pilon-Thomas, S., and Mule, J.J. (2008). Nonmyeloablative chemotherapy followed by T-cell adoptive transfer and dendritic cell-based vaccination results in rejection of established melanoma. *J Immunother* 31, 402–412.

70. de Witte, M.A., Jorritsma, A., Kaiser, A., van den Boom, M.D., Dokter, M., Bendle, G.M., Haanen, J.B., and Schumacher, T.N. (2008). Requirements for effective antitumor responses of TCR transduced T cells. *J Immunol* 181, 5128–5136.

71. Jiang, H.R., Gilham, D.E., Mulryan, K., Kirillova, N., Hawkins, R.E., and Stern, P.L. (2006). Combination of vaccination and chimeric receptor expressing T cells provides improved active therapy of tumors. *J Immunol* 177, 4288–4298.

72. Casati, A., Zimmermann, V.S., Benigni, F., Bertilaccio, M.T., Bellone, M., and Mondino, A. (2005). The immunogenicity of dendritic cell-based vaccines is not hampered by doxorubicin and melphalan administration. *J Immunol* 174, 3317–3325.

73. Apetoh, L., Ghiringhelli, F., Tesniere, A., Obeid, M., Ortiz, C., Criollo, A., Mignot, G., Maiuri, M.C., Ullrich, E., Saulnier, P., *et al.* (2007). Toll-like receptor 4-dependent contribution of the immune system to anticancer chemotherapy and radiotherapy. *Nat Med* 13, 1050–1059.

74. Kohlmeyer, J., Cron, M., Landsberg, J., Bald, T., Renn, M., Mikus, S., Bondong, S., Wikasari, D., Gaffal, E., Hartmann, G., *et al.* (2009). Complete regression of advanced primary and metastatic mouse melanomas following combination chemoimmunotherapy. *Cancer Res* 69, 6265–6274.

75. Salem, M.L., Kadima, A.N., El-Naggar, S.A., Rubinstein, M.P., Chen, Y., Gillanders, W.E., and Cole, D.J. (2007). Defining the ability of cyclophosphamide preconditioning to enhance the antigen-specific CD8+ T-cell response to peptide vaccination: Creation of a beneficial host microenvironment involving type I IFNs and myeloid cells. *J Immunother* 30, 40–53.

76. Teitz-Tennenbaum, S., Li, Q., Davis, M.A., Wilder-Romans, K., Hoff, J., Li, M., and Chang, A.E. (2009). Radiotherapy combined with intratumoral dendritic cell vaccination enhances the therapeutic efficacy of adoptive T-cell transfer. *J Immunother* 32, 602–612.

77. Verdeil, G., Marquardt, K., Surh, C.D., and Sherman, L.A. (2008). Adjuvants targeting innate and adaptive immunity synergize to enhance tumor immunotherapy. *Proc Natl Acad Sci USA* 105, 16683–16688.

78. Hodi, F.S., Butler, M., Oble, D.A., Seiden, M.V., Haluska, F.G., Kruse, A., Macrae, S., Nelson, M., Canning, C., Lowy, I., et al. (2008). Immunologic and clinical effects of antibody blockade of cytotoxic T lymphocyte-associated antigen 4 in previously vaccinated cancer patients. *Proc Natl Acad Sci USA* 105, 3005–3010.

79. Ribas, A., Comin-Anduix, B., Chmielowski, B., Jalil, J., de la Rocha, P., McCannel, T.A., Ochoa, M.T., Seja, E., Villanueva, A., Oseguera, D.K., et al. (2009). Dendritic cell vaccination combined with CTLA4 blockade in patients with metastatic melanoma. *Clin Cancer Res* 15, 6267–6276.

80. Capitini, C.M., Chisti, A.A., and Mackall, C.L. (2009). Modulating T-cell homeostasis with IL-7: Preclinical and clinical studies. *J Intern Med* 266, 141–153.

81. Sportes, C., Babb, R.R., Krumlauf, M.C., Hakim, F.T., Steinberg, S.M., Chow, C.K., Brown, M.R., Fleisher, T.A., Noel, P., Maric, I., et al. (2010). Phase I study of recombinant human interleukin-7 administration in subjects with refractory malignancy. *Clin Cancer Res* 16, 727–735.

82. Sportes, C., Gress, R.E., and Mackall, C.L. (2009). Perspective on potential clinical applications of recombinant human interleukin-7. *Ann N Y Acad Sci* 1182, 28–38.

83. King, J.W., Thomas, S., Corsi, F., Gao, L., Dina, R., Gillmore, R., Pigott, K., Kaisary, A., Stauss, H.J., and Waxman, J. (2009). IL15 can reverse the unresponsiveness of Wilms' tumor antigen-specific CTL in patients with prostate cancer. *Clin Cancer Res* 15, 1145–1154.

84. Berger, C., Berger, M., Hackman, R.C., Gough, M., Elliott, C., Jensen, M.C., and Riddell, S.R. (2009). Safety and immunologic effects of IL-15 administration in nonhuman primates. *Blood* 114, 2417–2426.

85. Miyagawa, F., Tagaya, Y., Kim, B.S., Patel, H.J., Ishida, K., Ohteki, T., Waldmann, T.A., and Katz, S.I. (2008). IL-15 serves as a costimulator in determining the activity of autoreactive CD8 T cells in an experimental mouse model of graft-versus-host-like disease. *J Immunol* 181, 1109–1119.

86. Davis, I.D., Brady, B., Kefford, R.F., Millward, M., Cebon, J., Skrumsager, B.K., Mouritzen, U., Hansen, L.T., Skak, K., Lundsgaard, D., *et al.* (2009). Clinical and biological efficacy of recombinant human interleukin-21 in patients with stage IV malignant melanoma without prior treatment: A phase IIa trial. *Clin Cancer Res* 15, 2123–2129.

87. Thompson, J.A., Curti, B.D., Redman, B.G., Bhatia, S., Weber, J.S., Agarwala, S.S., Sievers, E.L., Hughes, S.D., DeVries, T.A., and Hausman, D.F. (2008). Phase I study of recombinant interleukin-21 in patients with metastatic melanoma and renal cell carcinoma. *J Clin Oncol* 26, 2034–2039.

88. Rizzuto, G.A., Merghoub, T., Hirschhorn-Cymerman, D., Liu, C., Lesokhin, A.M., Sahawneh, D., Zhong, H., Panageas, K.S., Perales, M.A., Altan-Bonnet, G., *et al.* (2009). Self-antigen-specific CD8+ T cell precursor frequency determines the quality of the antitumor immune response. *J Exp Med* 206, 849–866.

89. Xue, S.A. and Stauss, H.J. (2007). Enhancing immune responses for cancer therapy. *Cell Mol Immunol* 4, 173–184.

90. Manzo, T., Hess Michelini, R., Basso, V., Ricupito, A., Chai, J.G., Simpson, E., Bellone, M., and Mondino, A. (2011). Concurrent allorecognition has a limited impact on posttransplant vaccination. *J Immunol* 186, 1361–1368.

Chapter 2

Gene Transfer into T Cells

David Gilham[1] and Hinrich Abken[2]

[1]*Clinical and Experimental Immunotherapy Group,*
School of Cancer and Enabling Sciences,
The University of Manchester,
Manchester Academic Healthcare Science Centre
[2]*Center for Molecular Medicine Cologne (CMMC), Tumorgenetics,*
University of Cologne, and Clinic I Internal Medicine,
Tumorgenetics, University Hospital Cologne,
Cologne, Germany

2.1 Introduction

The concept of adoptive T-cell immunotherapy is dependent upon the identification and expansion of antigen-specific T cells. Naturally occurring antigen-specific T cells are generally only present at a low frequency within the peripheral circulation (<1:1,000) which means that for therapeutic use, isolating these T cells is technically demanding and expanding this small population of cells to achieve the sort of numbers currently thought to be necessary for clinical application (>1 × 10⁷ cells/kg or a minimum of 10⁹ cells per adult patient) generally requires an extended period of *ex vivo* culture. Furthermore, in many cases, suitable T-cell antigens have not been described and, consequently, there may be no tools available to identify tumor-reactive T cells in the first place.

In order to tackle some of these issues, genetic modification has been explored as a route to generate clinically relevant numbers of antigen-specific T cells within a relatively short period of time. The two most commonly used approaches to date involve the expression of a recombinant T-cell receptor (TCR) and a chimeric antigen receptor (CAR), each of which is dealt with in detail in subsequent chapters. However, the key aims and issues relating to genetic modification are in general common for both approaches, and many are also relevant for other T-cell modification strategies, such as the expression of suicide genes in donor lymphocyte infusions for hematological malignancies.

The key prerequisite for tumor-specific receptor gene modification is that the altered T cell stably expresses the targeting receptor for a sufficient period of time to enable the T cell to interact with the tumor target and to initiate an immune-mediated response thereby leading to the eradication of the tumor. The length of time remains unclear; however, this period is likely to be in the order of weeks rather than days given that the modified T cell after infusion into the patient will need to travel to the site of tumor. There are other additional factors that the vector system used should deliver:

- High efficiency of gene transfer into primary T cells.
- High levels of vector stability with the transgene, the promoter, and associated elements delivered intact with no recombination occurring during vector production or after delivery to the primary T cell.
- Excellent safety profile.
- Have minimal deleterious effects upon the gene-modified T cell with respect to normal cellular biological functions.
- High potential for scale-up to produce sufficient vector titers to gene-modify large numbers of T cells for therapy and suitability for clinical application (i.e., that the vector system is able to comply with current regulations relating to the production of materials for clinical use).

Of the range of vectors available and tested for this approach, the family of γ-retroviral and lentiviral systems have been the most extensively used at the clinical level. As with most things, these vectors represent a compromise in that they can deliver most of the criteria listed above although with further fundamental understanding of T-cell biology and improved technology, there are now suspected drawbacks associated with these vectors. As such, the processes used to generate the retroviral gene-modified T cells are being improved and alternative systems that circumvent some of the issues associated with retroviral vectors are being developed.

2.2 γ-Retroviral Vectors to Generate Gene-Modified T Cells

The γ-retroviral vectors most commonly used have been based upon the type C *oncoretroviridae*, including the moloney murine leukosis virus (MLV).[1] Over the course of the last 20 years, these MLV vectors have been engineered to improve design, safety, and production with each of these areas reviewed in detail elsewhere.[2-5] In relation to the generation of gene-modified antigen-specific T cells, MLV vectors have proven worth due to the following reasons:

(i) Retroviral vectors can accommodate transgene cassettes in the region of 7–8 kb with multiple transgenes expressed using bicistronic expression elements including internal ribosome entry sites (IRES)[4] or viral proteolytic cleavage sequences such as the 2A sequence from the foot and mouth disease virus.[6] Consequently, retroviral vectors are suitable to express both CAR constructs and both α and β chains of the TCR potentially with selection markers to facilitate enrichment of the transduced T cells.

(ii) MLV-based vectors incorporating engineered elements to enhance and extend gene expression specifically in primary T cells have been generated and are currently in use in clinical trials (e.g., the pMP71 vector[7]).

(iii) Clinical-grade batches of retroviral vector of sufficient titer and volume have been produced and have been shown to maintain full activity when stored frozen over periods of years.[8]

The history of the clinical application of gene-modified T cells dates back to the late 1980's where T cells transduced with a neomycin resistance marker gene were safely infused into patients and were shown to infiltrate into tumor sites[9]. Subsequently, retroviral viral vectors encoding adenosine deaminase (ADA) were generated and were used to treat two patients in September 1990[10]. A long-term follow-up of these patients (12 years)[11] demonstrated that in one patient, 20% of circulating lymphocytes possessed the transgene while there was very low persistence (>0.1%) of gene-modified T cells in the second patient. Importantly, there was evidence of immunity generated against the T-cell product in the second patient which may explain the lack of persisting transduced T cells.[12] Moreover, a decade-long follow-up study of human immunodeficiency virus (HIV)-infected patients adoptively transferred with CD4z CAR-T cells confirmed persistence of these T cells with no evidence of pathological effect resulting from retroviral gene transfer[13]. Together, these early studies demonstrate that the retroviral gene-modification process is feasible at a clinical level and provided evidence of long-term gene expression *in vivo*. Furthermore, these studies suggested that γ-retroviral vectors were safe and that sufficient numbers of gene-modified T cells could be generated using these systems to support clinical trial application. Given this background, γ-retroviral vectors have proven to be the predominant, but not the only choice of vector for targeted T-cell clinical studies to date.[14–17]

2.3 The Potential Drawbacks of γ-Retroviral Vectors for T-Cell Gene Therapy

γ-Retroviral vectors suffer some potentially significant drawbacks: the three most critical being (i) shutdown of transgene expression through silencing of the retroviral promoter, (ii) that the viral vector stably integrates into the target cell genome thereby rising the potential risk of insertional mutagenesis, and (iii) the fact that efficient retroviral gene transfer is dependent upon the target cell undergoing cell division.[18]

Shutdown of the retroviral promoter is important since loss of the targeting receptor would prevent tumor recognition by the gene-modified T cell. Consequently, maintaining transgene expression at high level for as long as possible will be critical to allow the T-cell time to traffic to sites of tumor and interact with the target cell to drive an immune-based anti-tumor response. A recent trial of MART-1 TCR gene-modified T-cells using a Murine Stem Cell Virus (MSCV) long terminal repeat (LTR) and MFG vector backbone with α and β TCR chains co-expressed by means of an IRES element.[14] There was some evidence of reduced level of cell surface TCR expression at time points after 1 month as compared to the level of gene-marked circulating T cells suggesting some degree of LTR shutdown.[14] However, further investigation of T cells from these patients suggests that LTR shutdown was not specific and more reflected the general reduction in gene expression associated with the transition of the T cell from an activated to more quiescent state.[19] As such, it remains unclear whether vector shutdown may be specific for γ-retroviruses or may be a more general feature associated with the heterologous expression of genes within T-cells *in vivo*.

The probability of insertional mutagenesis driven by retroviral vectors was considered to be low until the cases of vector-related hyper-lymphoproliferation were observed in children treated with hemopoietic stem cells (HSCs) gene-modified with MLV vectors encoding the common γ chain to correct IL-2R deficiency in X-SCID.[20,21] Although subsequent preclinical models suggest that the expressed transgene was not oncogenic *per se*,[22] further preclinical studies have demonstrated insertional mutagenesis resulting from the adoptive transfer of gene-modified HSC.[23] Taken together, these studies raised serious concerns about the long-term safety of retroviral vectors which lead the regulatory bodies to question whether specific MLV retroviral vectors should be phased out from clinical use.[24] However, genotoxicity has not been identified in trials of gene-modified T cells carried out over the last two decades[25] while mouse studies suggest that mature T cells are resistant to transformation after transduction with retroviruses,[25] including studies using vectors encoding oncogenes; observations that are in contrast to that seen with the same vectors used to transduce CD34+ HSCs.[26] Furthermore, analysis of donor lymphocyte infusions genetically modified with the Herpes simplex virus-thymidine kinase (HSV-tk) gene appear to undergo clonal deletion rather than clonal expansion, arguing against an oncogenic effect associated with retroviral transduction of mature T cells.[27]

However, given this potential toxicity, efforts have been made to develop suicide systems which can specifically deplete gene-modified T cells *in vivo*. Donor lymphocyte infusions transduced with the HSV-tk transgene have been shown to clinically control graft versus host disease during the treatment of hematological malignancies; presumably, the mechanism of action being the specific depletion of autoreactive HSV-tk expressing T cells after administration of the pro-drug ganciclovir.[28–30]. In order to avoid the potential immunogenicity associated with the expression of viral transgenes in human T cells, fully humanized systems based upon FAS[31] or caspase-9[32] activation to drive controlled T-cell death have been recently developed. Taken together, the question of insertional mutagenesis as related to T-cell gene therapy remains unproven although, in anticipation of questions concerning the safety of these vectors, suicide gene systems are becoming increasingly developed and consequently, γ-retroviral vectors currently remain in favor for clinical T-cell gene therapy studies.

In contrast, the issue of cell division to facilitate retroviral gene transfer[18] has become of increasing importance to the field. In order to achieve efficient transduction of primary T cells, the isolated cells are given a strong mitogenic stimuli most commonly in the form of antibodies (e.g., anti-CD3ε, OKT3) or lectins (phytohemagluttinin, PHA) supported with interleukin-2 (IL-2) to drive active proliferation,[3,33,34] a process similar to that used to generate large numbers of natural antigen-specific human T cells without antigen presenting cells (APCs)[35]. After a 2–3 day stimulation period, the now rapidly cycling T cells are exposed to the retroviral vector. Simply mixing the retroviral supernatant with T cells generally results in low efficiency of gene transfer. As such, methods involving centrifugation of the viral supernatant with T cells ("spinfection"),[36] preloading of culture vessels by centrifugation of virus onto the surface of the vessel,[37] and co-localization of retrovirus and T cells on CH-296 ("Retronectin")-coated culture vessels.[38,39] Importantly, these methods have been transcribed to clinical production processes[33,40] thereby facilitating the large-scale production of gene-modified T cells. Following either single or multiple rounds of retroviral transduction, the T cells are then cultured for a period of time to achieve sufficient numbers of cells for application most usually for a period of up to 14 days, preferentially less than 10 days, and predominantly driven by the addition of IL-2. This generalized scheme outlines the three key aspects of retroviral gene transfer into primary human T cells — namely (i) T-cell activation, (ii) transduction, and (iii) expansion.

Whilst this process is relatively efficient at generating T cells within a regulated framework suitable for clinical application, the suitability for these T cells to function optimally *in vivo* is now being questioned at several levels. The gene-transduction process, subsequent culture, and selection using drug resistance markers where required, have been shown to adversely impact the potency of antiviral T-cell responses of gene-modified T cells as compared to control, non-transduced T-cells from the same donor[41] while also possibly skewing the range of viral-specific T-cell responses.[42] Additionally, the phenotype of the expanded T cells is altered toward a more differentiated phenotype likely to be suboptimal for adoptive transfer.[43,44] T-cell activation may also downregulate chemokine receptor expression thereby reducing the potential of these cells to traffic effectively *in vivo*; for instance, activation with anti-CD3 and anti-CD28 antibodies reduces the expression of CCR9 and $\alpha 4\beta 7$ integrin on primary T cells which would be predicted to impair trafficking to the small intestine.[45] Finally, the *ex vivo* expansion of T cells is likely to result in reduced telomere length thereby likely compromising the ability to survive in the long term *in vivo*. Work involving the adoptive transfer of *ex vivo* expanded tumor-infiltrating lymphocytes suggests that T cells that persist optimally after infusion into the patient should possess long telomeres and express cell surface markers including indicative of naïve/central memory phenotype such as CD27 and CD28.[46-48] The combination of these observations suggest that the current retroviral transduction and IL-2-driven expansion of T cells for therapy may prove to be compromised in terms of optimal *in vivo* performance as a direct result of the production process. In response to these issues, workers in the field have focused upon either improving upon the current technology or exploring novel approaches in order to produce gene-modified T cells that may be "fitter for *in vivo* function". These developments will be explored firstly in respect to improving the approach using γ-retroviral vectors and then to focus upon alternative approaches now being explored to produce T cells with a defined anti-tumor specificity.

2.4 Improving All Aspects of γ-Retroviral Gene Transfer to Preserve or Enhance Antigen-Specific T-Cell Function *In Vivo*

The three core aspects of retroviral gene modification, i.e., T-cell activation, transduction, and expansion, are all being actively investigated to determine what technical improvements can be introduced to improve the function and phenotype of the final T-cell product. The addition of anti-CD3ε monoclonal antibody with IL-2 to isolated peripheral blood mononuclear cells represents the most straightforward, clinically compatible system to activate T cells. However, the combination of an antibody against CD28 with an anti-CD3ε antibody appears to stimulate T cells has been shown to result in enhanced gene transfer, improved CD4$^+$ T-cell proliferation, reduced levels of activation-induced cell death (AICD), and also preserved to a degree T-cell alloreactivity.[49] Indeed, the method of transduction itself may impact upon the final T-cell phenotype; CH-296 recombinant fibronectin fragment ("Retronectin"), commonly used to facilitate retroviral transduction, was used alongside anti-CD3 antibody to make human T cells more susceptible to transduction with an eGFP marker retrovirus. This combination resulted in larger numbers of expanded gene-modified T cells which possessed a greater degree of cells expressing "naïve"-type CCR7$^+$ CD45RA$^+$ markers and engrafted in NOD/SCID mice to higher levels than T cells stimulated with anti-CD3 or anti-CD3/CD28 antibodies[50] suggesting the potentially important role of integrin signaling through retronectin when coupled with a mitogenic stimuli for the *ex vivo* expansion of T cells.

In order to provide a more reproducible presentation of the antibodies to T cells, paramagnetic beads coated with anti-CD3/anti-CD28 antibodies have been used to drive T-cell expansion.[49] These beads, called artificial antigen presenting cells (aAPC),[51] have been used clinically to expand donor lymphocyte infusion (DLI)[52] and for retroviral transduction with HSV-tk co-expressed with the truncated nerve growth factor receptor (tNGFR).[53] These beads are now also commercially available in slightly differing formats (e.g., Dynal/Invitrogen, Miltenyi Biotec) and are currently being used to generate gene-modified antigen-specific T cells for clinical trial. The bead-based format has other advantages including providing the potential for magnetic-based cell separation and also flexibility in that the concentration of anti-CD3/CD28 antibodies can be varied as required for optimal T-cell subset expansion while also allowing differing protein products to be bound to the bead, e.g., HLA-tetramers in place of the anti-CD3 antibody for the specific activation of antigen-defined T-cell populations.[54]

Once more, technology moves on and the structural role of the bead component of the aAPC itself has been further examined. Clustering of anti-CD3/anti-CD28/LFA-1 on microdomains presented on GM-1-enriched liposome scaffolds have been used to mimic the fluid nature of a cellular membrane resulting in an improved expansion of gene-modified MART-1-specific T cells.[55] However, better than mimicking the cell membrane is to use one. To this end, cell-based aAPC's have been developed where a suitable cell line is modified to express ligands suitable to drive T-cell expansion. NIH3T3 cells transduced with five vectors expressing human HLA-A2.1, β2-microglobulin, ICAM-1, LFA-3, and CD80 proved to be suitable stimulators to drive the expansion of CD8$^+$ HLA-A2.1 restricted T cells.[56] However, the erytho-leukamic cell line K562 has been further explored as the basis for an aAPC cell line. There are many potential advantages of using the K562 system including the fact that it is a non-adherent cell line so ideal for bioreactor culture systems. Further, K562 transfected with constructs encoding Fc receptors allow these cells to be effectively coated with antibodies of choice, including anti-CD3 and anti-CD28 antibodies.[57,58] Recent work suggests that a strong CD28 co-stimulation, provided in the context of the K562 aAPC, was essential to expand and maintain the function of human regulatory T (Treg) cells.[59] Whilst potentially attractive for the expansion of Treg's for autoimmune indications, avoiding the expansion of this T-cell subset is likely to be of major importance for the expansion of tumor-specific T cells. To this end, stimulation of 4–1BB expressed on T cells with K562 expressing 4–1BBL appears to generate memory CD8[+] T cells that are superior in quality for adoptive therapy as compared to those generated using CD28 co-stimulation.[57,60] Taken together, these studies confirm the flexibility of the aAPC system to manipulate the phenotype of the T-cell product with clinical testing of these T cells now underway to determine whether these preclinical observations carry through to a clinical readout such as improved engraftment and survival of the adoptively transferred T cells.

Cytokines play a central role in driving the expansion and survival of cultured T cells, with IL-2 being the key cytokine used in the majority of gene-modification protocols. However, there is an increasing degree of literature suggesting that various cytokines and combinations of cytokines including IL-7 with IL-21,[61] IL-15 with IL-21,[62,63] and IL-18[64] can improve the engraftment of human T cells in *in vivo* models and stimulate increased anti-tumor function of cultured T cells. In terms of CAR T cells, the addition of IL-12 to anti-CD3 antibody and IL-2 stimulation to generate carcino-embryonic antigen (CEA)-specific T cells resulted in a reduction of Th2 cytokines in the activated T cells and improved the potency of the T cells to eliminate a short-term established tumor in nude mice.[65] In a hematological model, IL-15 combined with fibroblast aAPC's expressing CD80 improved the eradication of 6- or 7-day established systemic Raji B-lymphoma cells by CD19-specific CAR T cells.[66] These early studies suggest that the cytokines used during T-cell activation and expansion are likely to have a major impact upon the function of the final T-cell product. However, the optimal cytokines or combination of cytokines is not yet understood and for final clinical development, the availability of individual cytokines to clinical standard may represent a limiting factor in the full translation of these preclinical studies.

Taken together, these advances in T-cell activation and culture conditions are overcoming some of the drawbacks associated with the use of γ-retroviral vectors. However, these protocol-based modifications do not circumvent the central issue associated with the requirement to pre-activate T cells for retroviral transduction. Given this, researchers have actively explored alternative viral and non-viral systems to determine whether high efficiency gene transfer into unstimulated T cells is feasible.

2.5 Lentiviral Vectors to Transduce Minimally Stimulated and Quiescent Primary Human T Cells

Lentiviruses, like γ-retroviruses, are members of the *retroviridae* RNA-containing virus family. There are several significant structural and genetic differences between the two subfamilies: of most importance here is the fact that lentiviruses possess additional accessory genes which confer the ability to stably transduce non-dividing cells. Indeed, engineering of the HIV-1 lentivirus has generated a vector capable of transducing quiescent cells.[67] Nonetheless, quiescent primary human T cells are not permissive to wild-type HIV-1 replication or transduction with HIV-1 vectors.[68] However, human T cells stimulated with mitogenic stimuli (in a manner similar to that used with γ-retroviral vectors) are potentially more permissive to transduction by HIV-1 vectors than γ-retroviral vectors.[68-72] Moreover, HIV-1 vectors can also transduce human T cells that are cultured in the presence of cytokines without mitogenic stimulation; a situation where MLV vectors are poorly effective.[68-72] There has been some debate concerning the exact role of the HIV-1 accessory proteins with one group suggesting that these accessory proteins are required for the efficient transduction of cytokine-stimulated T cells[70] while others argue against this.[72] Certain structural elements of the HIV-1 vector appear to be specifically important for the efficient transduction of T lymphocytes and monocytes including the maintenance of the central polypurine tract (cPPT). However, it is clear that T-cells stimulated with common γ-chain cytokines including IL-7 and IL-2 can be efficiently transduced with HIV-1 lentiviral vectors and that under these conditions, the T cells produced display an increased preserved immune competence and express cell surface markers indicative of reduced differentiation.[68–72]

The attraction of lentiviral vectors is therefore the potential ability to generate cytokine-stimulated gene-modified T cells which may be predicted to be superior in phenotype and function as compared to those generated using γ-retroviral vectors. However, to date, this has not been definitively proven to be the case. Lentiviral vectors encoding CAR's have been shown to transduce and express the receptor efficiently but only in anti-CD3 antibody-stimulated T cells.[73,74] For TCR's, the expression of α and β chains of a melanoma-specific TCR in the context of a single lentiviral vector has been optimized in terms of developing 2A proteo-cleavage sequences linked with furan cleavage sites[75] and promoters.[76] However, clinical-scale transduction of peripheral blood lymphocytes using a HIV-1 vector encoding the F5 MART-1-specific TCR utilized anti-CD3/CD28 stimulation to expand the transduced T cells to sufficient numbers for clinical use (3×10^{10} total lymphocytes) although there was a suggestion that the expanded T cells maintained a relatively increased frequency of less-differentiated CD45RO$^-$ CD62L$^+$ transduced T cells.[77] More recently, lentiviral vectors encoding a CD19-specific CAR have been used to transduce T cells used to treat three patients with chronic lymphocytic leukemia.[78,79] The impressive clinical responses seen in these three patients clearly support the CAR-T cell concept. However, it remains unclear whether the lentiviral vector itself was providing a significant advantage to the overall therapy in this study since, among other issues, the CAR used was a new variant meaning that it is difficult to tease apart the potential contributions of each part of the therapy. Nonetheless, this recent study clearly demonstrates the potential for CAR-T cells and supports the potential of lentiviral vectors as agents to generate gene-modified T cells for clinical application.

A recent potentially important development has been the pseudotyping of a HIV-1 vector with measles virus glycoproteins replacing the more commonly used vesicular stomatitis virus G (VSV-g) envelope protein. Initial reports indicate that transduction levels of

50% were achieved in quiescent T cells in the absence of cytokine stimulation and cell cycling.[80] The mechanism of viral transduction by measles virus pseudotyped lentiviral vectors is currently not known; however, the measles virus envelope is suitable for further engineering and may permit more specific retargeting of the virus to cellular subtypes and also potentially increased titers compared to those achieved with VSV-g pseudotyped lentiviral vectors.[81,82] Additionally, clinical trials using HIV-1 vectors have now been initiated demonstrating the feasibility of the lentiviral approach to generate gene-modified T-cells, in this case CD4$^+$ cells expressing an anti-sense RNA construct directed against the envelope protein of HIV.[83] As such, the reagents and processes are becoming available to examine the relative merits of lentiviral transduction compared to γ-retroviral transduced T cells. Aside from the potential to transduce minimally stimulated or quiescent primary T cells, there have been strong suggestions in the field of HSC transfer that lentiviral vectors offer an enhanced degree of safety from insertional mutagenesis than that of γ-retroviruses with specific reference to the case of X-linked SCID.[84–86] However, there has been a very recent and preliminary report of clonal dominance potentially as a result of lentivector integration into the locus for high mobility group-A2 (HMW-A2) proteins seen in a trial of β-globin gene transfer into CD34$^+$ HSC's.[87] This observation is clearly made at an early stage; nonetheless, this single observation may suggest that there may be issues potentially residing with integrating vectors *per se* and not just the γ-retroviral vectors.

2.6 Other Viral Vectors Used to Transduce Primary Human T Cells

The γ-retroviral and lentiviral vector families have been the most extensively used to generate gene-modified and antigen-specific T cells largely based upon improved protocols that generate high efficiency of transduction and stable integration thereby allowing a relatively small number of T cells to be transduced and then expanded to the required cell numbers. Other viral vectors remain far less developed in terms of this approach although there are some candidates which offer their own potential advantages. Foamy virus vectors are *retroviridae* family members (genus: *spumavirus*) and thought to possess little or no pathogenic activity in humans which is a highly desirable feature for any potential gene transfer vector.[88] There is one report documenting the efficient transduction (> 80%) of IL-2 stimulated peripheral blood lymphocytes and PHA stimulated lymphocytes with a Lac-Z marker gene encoding simian foamy virus suggesting that T cells are permissive to this vector.[89]

As an alternative to RNA viruses, adenoviral vectors have been shown to have the capacity to transduce primary T cells although the most commonly used serotype Ad5 transduces this cell type relatively inefficiently. Transduction efficiencies can be improved by using bispecific antibodies to target the virus to CD3[90] or by coating the virus in lipid complexes.[91] Alternatively, different adenoviral serotypes and particularly Ad35 appears to transduce primary T cells to higher efficiency than the Ad5 serotype.[92,93] The predominant advantage of the adenoviral vector is the ability to produce extremely high titers which is important from a biomanufacturing point of view and also may transduce minimally stimulated T cells. However, the virus does not integrate into the host genome and, consequently, the expansion of T cells *ex vivo* is not possible without loss of the vector. Whether such a system will be applicable to the production of antigen-specific T cells remains uncertain. However, another DNA virus, adeno-associated virus (AAV) encoding a GFP marker gene has been directly injected into the thymus of test macaques and shown to result in the presence of gene-modified lymphocytes within secondary lymphoid organs 10 days after gene transfer with cells remaining at detectable levels until day 30.[94]

At present, these alternative gene transfer systems have been tested using marker genes in proof of principle studies to confirm the permissiveness of T cells to transduction. Although as yet untested, the increasing battery of vectors will most likely provide different transduction conditions which may have either positive or negative effects upon the *in vitro* and *in vivo* function of antigen-specific T cells.

2.7 Non-Viral Gene Transfer Into Primary Human T Cells: Plasmid-Based Systems

The high transduction efficiency of viral vectors is balanced by some serious practical issues; retroviral and lentiviral vectors can only accommodate a relatively limited transgene capacity while the production and storage of the clinical-scale vector, along with the generation of the viral-vector transduced T cells, is technically demanding requiring specialized facilities and challenging in terms of achieving regulatory approval meaning that the whole process is generally costly. The manufacture of large-scale plasmid DNA is a cheaper option and as such is attractive to generate gene-modified T cells. The gene transfer approach is also more straightforward employing electroporation of anti-CD3 activated peripheral blood lymphocytes.[95,96] However, the drawback of plasmid DNA transfer remains the low level of stable transgene expression since the frequency of integration of the plasmid vector into the host T cell genome is low[96] even when using more recent nucleofection technology.[97] To compensate for this, the plasmid vectors used to express the CAR also include drug resistance genes including neomycin phosphotransferase (*neo*), hygromycin phosphotransferase (*hygro*) and thymidine kinase (HSV-tk). The electroporated T cells are cultured in the presence of the relevant drug for a period of time before being washed free of drug and subsequently expanded. The transfected T cells may be cloned[96] or expanded as a bulk culture[95]; the downside of this method is the relatively long culture period which can be 42 days or more depending upon the rate of proliferation of the T-cell product although the use of K562 cells in the context of an aAPC appeared to improve the degree of expansion of CD8+ CAR T cells.[98] As an alternative approach, immunomagnetic selection of transfected T cells expressing a cell surface truncated LNGFR gene has been shown to allow the selection of stable transfected CD4+ T cells, but this method again involved a lengthy period of culture and multiple selections to achieve a stable population.[97]

Importantly, this approach has been used in several phase I clinical trials targeting neuroblastoma (Li-CAM specific),[99] non-Hodgkin's lymphoma (CD20),[100] and glioma (IL-13 zetakine).[101] Gene-modified T cells could be detected for varying periods in both the neuroblastoma and lymphoma trials but there were some suggestion that the cytoreductive chemotherapy given prior to T-cell infusion may have enhanced T-cell persistence.[100] There was also no evidence of immunological responses against the infused T cells found in the lymphoma patients despite the expression of bacterial or viral drug selection markers although, once again, the cytoreductive therapy used may have played a role in limiting the immune response in this situation.

2.8 Non-Viral Gene Transfer Into Primary Human T Cells: Transposon Technology

A significant recent development of the plasmid-based approach has been transposon technology. Sleeping Beauty has been derived from the Tc1/mariner superfamily of transposons and consists of the transposable element from a DNA donor plasmid adapted for non-viral gene transfer in T cells using a Sleeping Beauty transposase supplied in trans to mediate integration of the transposon CAR expression cassette flanked by terminal inverted repeats, which each contain two copies of a short direct repeat with binding sites for the transposase enzyme. The latter mediates transposition by binding to the inverted repeats, excising the DNA flanked by the inverted repeats and inserting the transposon into any of about 200 million TA sites in the mammalian genome.[102,103] Transfer of the donor vector and Sleeping Beauty transposase is accomplished using electroporation or nucleofection; as a result, quiescent T cells are readily transfected using this system[104] and these cells can then be expanded using rapid expansion protocols to achieve large numbers of T cells. Sleeping Beauty has been used in conjunction with a CD19-specific CAR with CD4+ and CD8+ T-cell subsets equally transfected and when combined with the K562 aAPC expansion system[104] or CD3/CD28 beads,[105] the levels of CD19+ CAR T cells are maintained indicating stable expression of the T cells. Sleeping Beauty has also been used with a melanoma-specific TCR[106] and with CD19-specific CAR's[104,105,107,108] while other transposon systems with potentially increased transgene carrying ability are being developed for T-cell transfection and CAR expression.[109,110] Overall, the transposon gene-transfer system is highly attractive given the potentially reduced costs of vector manufacture and storage; comparisons with viral gene-modified T cells are clearly warranted.

2.9 Non-Viral Gene Transfer Into Primary Human T Cells: RNA Vectors

T cells can be gene modified by mRNA electroporation coding for the respective receptor without integration-associated safety concerns, however, with transient expression. RNA transfer was used to transfer full-length TCR and, most recently, CARs to T cells in order to redirect them with a new specificity.[111-116] RNA, generated by *in vitro* transcription from DNA cloned into special vectors or from polymerase chain reaction generated, vector-free DNA templates, is transferred into T cells by electroporation. While the procedure to express CAR encoding mRNA in T cells was repeatedly described in the last years, Birkholz *et al.* recently compared redirected effector functions of human T cells *in vitro* engineered by either RNA or retroviral transfer.[111] Human CD4+ and CD8+ T cells from peripheral blood can be engineered by RNA transfer with high efficiencies, i.e., >90%, to express the CAR on cell surface. CAR expression, however, is transient, although constant for about 24 h after electroporation, thereafter decreasing to half-maximal expression at day 2 and no detectable expression at day 9 after electroporation.[111,117] Engineered T cells are susceptible to CAR-triggered activation only during this time and loss of CAR expression results in loss of redirected specificity. T cells reprogrammed by RNA transfer are fully functional *in vitro*, resulting in CAR-triggered cytokine production and specific tumor cell lysis. For clinical applications, it is a prerequisite that reprogrammed T cells reach and enter the tumor where they retain and execute their lytic activity for a sufficient period of time. While retroviral modification of T cells allows constitutive expression of the transgenic TCR or CAR for at least several months, RNA transfer results in only transient CAR expression. Long-term killing assays *in vitro* revealed that

RNA-transfected T cells retained their cytotoxic function after 2 days of activation and exhibited cytolytic activities as retrovirally transduced T cells although at lower levels.[111]

RNA modified human T-cells-mediated regression of large vascularized mesotheliomas in immune-deficient mice; multiple T-cell injections, however, were required to produce lasting tumor regression.[117] A single injection of anti-CD19 RNA-modified T cells reduced xenografted leukemia in immunodeficient mice within 1 day after administration.[118]

There are some aspects which additionally need to be considered in context of clinical use. The procedure of RNA transfection is very fast in comparison to retroviral gene transfer and results in high numbers of modified T cells, i.e., more than 80% of T cells expressed the transgenic CAR after electroporation, rendering selection and excessive *ex vivo* expansion of engineered T cells unnecessary. Since up to 2×10^9 CD8$^+$ T cells can be generated from a lymphapheresis, sufficient numbers of modified T cells can be generated *ex vivo*. A high throughput microelectroporation device for RNA modification of peripheral blood T cells has recently been developed,[119] other devices are currently under construction. The electroporation of such quantities of T cells seems to be technically feasible.

Due to the fact that RNA does not integrate into the genome and is quickly degraded, this technique has certain advantages over retroviral transduction. Although neoplastic transformation of mature T cells as a result of a retroviral gene transfer is only rarely observed in experimental systems so far; at least theoretically, integration of the provirus into the genome bears the risk of insertional mutagenesis and malignant transformation of engineered T cells which is not expected upon RNA transfer.

RNA-modified T cells provide a flexible platform for cancer treatment that may complement the use of retroviral and lentiviral-engineered T cells, in particular when a CAR with unknown side effects is applied for the first time in humans. Stable expression of the transgene as achieved by retro- and lentiviral gene transfer may be a disadvantage in some clinical situations. Cross-reactivity of the CAR with healthy tissues, in particular, results in autoimmunity up to severe auto-aggression which is not self-limiting when modified T cells with stable CAR expression were applied. This was observed in a trial in which turned out that the target antigen was not only expressed on carcinoma cells but also on bile duct epithelium which resulted in grade 2–4 liver toxicity.[17] Although the clinical situation could be solved by application of high-dose steroids, there is obviously a need to explore auto-aggression by first applications of transiently modified T-cells. Although RNA-modified T cells may also produce acute toxicities, transient CAR expression does not allow extensive CAR-triggered expansion of transfected T cells at the targeted tumor site which is frequently considered as a disadvantage compared to retroviral transduction. In case of unintended autoimmune reactions, however, transient CAR expression may be a significant advantage. As a consequence, on the other hand, adoptive immunotherapy using RNA-engineered T cells will require repetitive applications, and no long-lasting memory will be established. Moreover, the CAR surface expression may be titrated by RNA transfer, giving T cells with potentially tunable levels of effector functions.[118] To combine the advantages and overcome limitations, RNA and retrovirally-modified T cells may be applied in a two-step clinical procedure. The initial testing of new CARs or TCRs may be done by RNA-electroporated T cells with transient CAR expression checking for potential acute autoimmunity which would rule out the tested receptor. If suitable in this respect, T cells with permanent CAR expression engineered by retroviral transduction may be applied in a second step in order to take advantage of the benefits of a long-lasting immune response and generation of specific memory. Since RNA transfection does not transfer gene(s) and does not result in permanent genetic alterations, regulatory hurdles to apply RNA transfection in clinical trials are expected to be much lower compared to those of retroviral transduction.

Taken together CAR expression upon RNA transfer is transient, providing a strategy to generate large numbers of tumor-specific, safe, and self-limiting T lymphocytes for the immunotherapy of cancer.

2.10 Mouse Models of Gene-Modified T-Cell Therapy

Mouse models have been the mainstay research route to test gene-modified T-cell therapies. Importantly, the principles laid for human T-cell gene transfer relate closely to the mouse meaning that the mouse T cell is a realistic subject for preclinical studies of adoptive T-cell transfer. Mouse T cells are sensitive to the mode of activation[120] and pseudotyping of the retrovirus.[121] However, protocols that detail specific methods to achieve high level, reproducible genetic modification of mouse T cells using retroviral vectors are available.[122–124] Like human T cells, the cytokines used to culture T cells after transduction are also likely to have a major impact upon T-cell function and, ultimately, therapy. Studies using mouse T cells bearing from the transgenic mice expressing a TCR specific for gp100 (pmel-1) have shown that T-cell dose and differentiation status have an effect upon *in vivo* anti-tumor efficacy.[125] Moreover, γc cytokines other than IL-2 (i.e., IL-7, IL-15, and IL-21) each appeared to support the production of T cells with improved anti-tumor activity as compared to T cells cultured in IL-2.[125] Indeed, mouse T cells transduced with retroviruses encoding a TCR also showed enhanced *in vitro* activity when cultured in IL-15 and IL-21.[126] Further studies to understand the impact of cytokines upon the *ex vivo* culture of gene-modified mouse T cells are clearly warranted.

Lentiviral vectors have been less commonly reported in mouse T-cell gene-transfer studies. Recent studies have confirmed that mouse T cells are suitable targets for lentiviral gene transfer using HIV-1-based vectors.[127] However, more recent work indicates that lentiviral vector-driven expression of TCR's is somewhat reduced compared to that achieved by γ-retroviral vectors expressing the same TCR,[123] which questions whether lentiviral vectors offer any significant advantage in studies employing mouse T cells.

2.11 Summary

Primary T-cell gene-transfer technology has made considerable progress over the last 20 years. Initial trials using gene marking started in the 1990's have shown the approach to be feasible and, importantly, to date there have been no long-term adverse effects reported that relate to toxicity of the vectors used. In the late 1990's, retroviral technology and T-cell culture methods had developed to permit routinely high levels of gene transfer[128] allowing a focus upon engineering of the CAR and TCR therapeutic unit and the development of protocols that meet current criteria for the generation of medicinal products, including the requirement of all aspects of the therapeutic product and the vector to comply with current good manufacturing process.

With clinical trials of TCR, CAR, and other gene-modified T-cell products either underway or due to start,[129,130] there has been a shift in focus with many researchers investigating more basic aspects of T-cell biology in order to identify a T cell which may have properties that offer the potential for improved therapeutic activity. In particular, less differentiated T cells from the naïve and central memory circulating pool[125,131–133] are thought to offer better anti-tumor activity. Indeed, the recent description of memory T cell that possess stem-cell like properties may further identify a specific T-cell subset that could offer many advantages to adoptive T-cell therapy over the use of total T-cell populations.[134] Consequently, with further refinements in the types of T cells that are being used for adoptive T-cell therapy, the future challenge will be to ensure that high levels of gene transfer into these specific T-cell subsets can be achieved with minimal deleterious impact upon the T cell itself.

References

1. Vile, R.G. and Russell, S.J. (1995). Retroviruses as vectors. 51, 12–30.
2. Baum, C., Schambach, A., Bohne, J., and Galla, M. (2006). Retrovirus vectors: Toward the plentivirus? 13, 1050–1063.
3. Hombach, A., Heuser, C., and Abken, H. (2003). Generation, expression, and monitoring of recombinant immune receptors for use in cellular immunotherapy. 207, 365–381.
4. Morgan, R.A. *et al.* (1992). Retroviral vectors containing putative internal ribosome entry sites: Development of a polycistronic gene transfer system and applications to human gene therapy. 20, 1293–1299.

5. Sinn, P.L., Sauter, S.L., and McCray, Jr., P.B. (2005). Gene therapy progress and prospects: Development of improved lentiviral and retroviral vectors — Design, biosafety, and production. 12, 1089–1098.

6. Klump, H. *et al.* (2001). Retroviral vector-mediated expression of HoxB4 in hematopoietic cells using a novel coexpression strategy. 8, 811–817.

7. Engels, B. *et al.* (2003). Retroviral vectors for high-level transgene expression in T lymphocytes. 14, 1155–1168.

8. Lamers, C.H. *et al.* (2008). Retroviral vectors for clinical immunogene therapy are stable for up to 9 years. 15, 268–274.

9. Rosenberg, S.A. *et al.* (1990). Gene transfer into humans — Immunotherapy of patients with advanced melanoma, using tumor-infiltrating lymphocytes modified by retroviral gene transduction. 323, 570–578.

10. Blaese, R.M. *et al.* (1995). T lymphocyte-directed gene therapy for ADA-SCID: Initial trial results after 4 years. 270, 475–480.

11. Muul, L.M. *et al.* (2003). Persistence and expression of the adenosine deaminase gene for 12 years and immune reaction to gene transfer components: Long-term results of the first clinical gene therapy trial. 101, 2563–2569.

12. Tuschong, L., Soenen, S.L., Blaese, R.M., Candotti, F., and Muul, L.M. (2002). Immune response to fetal calf serum by two adenosine deaminase-deficient patients after T cell gene therapy. 13, 1605–1610.

13. Scholler, J. *et al.* (2012). Decade-long safety and function of retroviral-modified chimeric antigen receptor T cells. *Sci Transl Med* 4, 132ra53.

14. Morgan, R.A. *et al.* (2006). Cancer regression in patients after transfer of genetically engineered lymphocytes. 314, 126–129.

15. Kershaw, M.H. *et al.* (2006). A phase I study on adoptive immunotherapy using gene-modified T cells for ovarian cancer. 12, 6106–115.

16. Mitsuyasu, R.T. *et al.* (2000). Prolonged survival and tissue trafficking following adoptive transfer of CD4zeta gene-modified autologous CD4(+) and CD8(+) T cells in human immunodeficiency virus-infected subjects. 96, 785–793.

17. Lamers, C.H. *et al.* (2006). Treatment of metastatic renal cell carcinoma with autologous T-lymphocytes genetically retargeted against carbonic anhydrase IX: First clinical experience. *J Clin Oncol* 24, e20–e22.

18. Miller, D.G., Adam, M.A., and Miller, A.D. (1990). Gene transfer by retrovirus vectors occurs only in cells that are actively replicating at the time of infection. 10, 4239–4242.

19. Burns, W.R., Zheng, Z., Rosenberg, S.A., and Morgan, R.A. (2009). Lack of specific gamma-retroviral vector long terminal repeat promoter silencing in patients receiving genetically engineered lymphocytes and activation upon lymphocyte restimulation. *Blood* 114, 2888–2899.

20. Hacein-Bey-Abina, S. *et al.* (2003). A serious adverse event after successful gene therapy for X-linked severe combined immunodeficiency. 348, 255–256.

21. Hacein-Bey-Abina, S. *et al.* (2003). LMO2-associated clonal T cell proliferation in two patients after gene therapy for SCID-X1. 302, 415–419.

22. Thrasher, A.J. *et al.* (2006). Gene therapy: X-SCID transgene leukaemogenicity. 443, E5–E6, discussion E6–7.

23. Li, Z. *et al.* (2002). Murine leukemia induced by retroviral gene marking. 296, 497.

24. Buchholz, C.J. and Cichutek, K. (2006). Is it going to be SIN? A European Society of Gene Therapy commentary. Phasing-out the clinical use of non self-inactivating murine leukemia virus vectors: Initiative on hold. 8, 1274–1276.

25. Westwood, J.A. *et al.* (2008). Absence of retroviral vector-mediated transformation of gene-modified T cells after long-term engraftment in mice. *Gene Ther* 15, 1056–1066.

26. Newrzela, S. *et al.* (2008). Resistance of mature T cells to oncogene transformation. *Blood* 112, 2278–2286.

27. Recchia, A. *et al.* (2006). Retroviral vector integration deregulates gene expression but has no consequence on the biology and function of transplanted T cells. *Proc Natl Acad Sci USA* 103, 1457–1462.

28. Bonini, C. *et al.* (1997). HSV-TK gene transfer into donor lymphocytes for control of allogeneic graft-versus-leukemia. *Science* 276, 1719–1724.

29. Ciceri, F. *et al.* (2009). Infusion of suicide-gene-engineered donor lymphocytes after family haploidentical haemopoietic stem-cell transplantation for leukaemia (the TK007 trial): A non-randomised phase I-II study. *Lancet Oncol* 10, 489–500.

30. Hollatz, G. *et al.* (2008). T cells for suicide gene therapy: Activation, functionality and clinical relevance. *J Immunol Methods* 331, 69–81.

31. Thomis, D.C. *et al.* (2001). A fas-based suicide switch in human T cells for the treatment of graft-versus-host disease. *Blood* 97, 1249–1257.

32. Straathof, K.C. *et al.* (2005). An inducible caspase 9 safety switch for T-cell therapy. *Blood* 105, 4247–4254.

33. Lamers, C.H., Willemsen, R.A., Luider, B.A., Debets, R., and Bolhuis, R.L. (2002). Protocol for gene transduction and expansion of human T lymphocytes for clinical immunogene therapy of cancer. *Cancer Gene Ther* 9, 613–263.

34. Riviere, I., Gallardo, H.F., Hagani, A.B., and Sadelain, M. (2000). Retroviral-mediated gene transfer in primary murine and human T-lymphocytes. *Mol Biotechnol* 15, 133–142.

35. Riddell, S.R. and Greenberg, P.D. (1990). The use of anti-CD3 and anti-CD28 monoclonal antibodies to clone and expand human antigen-specific T cells. *J Immunol Methods* 128, 189–201.

36. Gilham, D.E. *et al.* (2002). Primary polyclonal human T lymphocytes targeted to carcino-embryonic antigens and neural cell adhesion molecule tumor antigens by CD3zeta-based chimeric immune receptors. *J Immunother* 25, 139–151.

37. Kuhlcke, K. *et al.* (2002). Highly efficient retroviral gene transfer based on centrifugation-mediated vector preloading of tissue culture vessels. *Mol Ther* 5, 473–478.

38. Hanenberg, H. *et al.* (1996). Colocalization of retrovirus and target cells on specific fibronectin fragments increases genetic transduction of mammalian cells. *Nat Med* 2, 876–882.

39. Pollok, K.E. *et al.* (1998). High-efficiency gene transfer into normal and adenosine deaminase-deficient T lymphocytes is mediated by transduction on recombinant fibronectin fragments. *J Virol* 72, 4882–4892.

40. Lamers, C.H. *et al.* (2008). Retronectin-assisted retroviral transduction of primary human T lymphocytes under good manufacturing practice conditions: Tissue culture bag critically determines cell yield. *Cytotherapy* 10, 406–416.

41. Sauce, D. *et al.* (2002). Retrovirus-mediated gene transfer in primary T lymphocytes impairs their anti-Epstein-Barr virus potential through both culture-dependent and selection process-dependent mechanisms. *Blood* 99, 1165–1173.

42. Sauce, D. *et al.* (2003). Retrovirus-mediated gene transfer in polyclonal T cells results in lower apoptosis and enhanced *ex vivo* cell expansion of CMV-reactive CD8 T cells as compared with EBV-reactive CD8 T cells. *Blood* 102, 1241–1248.

43. Lamana, M.L., Bueren, J.A., Vicario, J.L. and Balas, A. (2004). Functional and phenotypic variations in human T cells subjected to retroviral-mediated gene transfer. *Gene Ther* 11, 474–482.

44. Duarte, R.F. *et al.* (2002). Functional impairment of human T-lymphocytes following PHA-induced expansion and retroviral transduction: Implications for gene therapy. *Gene Ther* 9, 1359–1368.

45. Eksteen, B. *et al.* (2009). Gut homing receptors on CD8 T cells are retinoic acid dependent and not maintained by liver dendritic or stellate cells. *Gastroenterology* 137, 320–329.

46. Tran, K.Q. *et al.* (2008). Minimally cultured tumor-infiltrating lymphocytes display optimal characteristics for adoptive cell therapy. *J Immunother* 31, 742–751.

47. Shen, X. *et al.* (2007). Persistence of tumor infiltrating lymphocytes in adoptive immunotherapy correlates with telomere length. *J Immunother* 30, 123–129.

48. Zhou, J. *et al.* (2005). Telomere length of transferred lymphocytes correlates with *in vivo* persistence and tumor regression in melanoma patients receiving cell transfer therapy. *J Immunol* 175, 7046–7052.

49. Kalamasz, D. *et al.* (2004). Optimization of human T-cell expansion *ex vivo* using magnetic beads conjugated with anti-CD3 and Anti-CD28 antibodies. *J Immunother* 27, 405–418.

50. Yu, S.S. *et al.* (2008). *In vivo* persistence of genetically modified T cells generated *ex vivo* using the fibronectin CH296 stimulation method. *Cancer Gene Ther* 15, 508–516.

51. Kim, J.V., Latouche, J. B., Riviere, I., and Sadelain, M. (2004). The ABCs of artificial antigen presentation. *Nat Biotechnol* 22, 403–410.

52. Porter, D.L. *et al.* (2006). A phase 1 trial of donor lymphocyte infusions expanded and activated *ex vivo* via CD3/CD28 costimulation. *Blood* 107, 1325–1331.

53. Orchard, P.J. *et al.* (2002). Clinical-scale selection of anti-CD3/CD28-activated T cells after transduction with a retroviral vector expressing herpes simplex virus thymidine kinase and truncated nerve growth factor receptor. *Hum Gene Ther* 13, 979–988.

54. Maus, M.V., Riley, J.L., Kwok, W.W., Nepom, G.T., and June, C.H. (2003). HLA tetramer-based artificial antigen-presenting cells for stimulation of CD4+ T cells. *Clin Immunol* 106, 16–22.

55. Zappasodi, R. *et al.* (2008).The effect of artificial antigen-presenting cells with preclustered anti-CD28/-CD3/-LFA-1 monoclonal antibodies on the induction of *ex vivo* expansion of functional human antitumor T cells. *Haematologica* 93, 1523–1534.

56. Latouche, J.B. and Sadelain, M. (2000). Induction of human cytotoxic T lymphocytes by artificial antigen-presenting cells. *Nat Biotechnol* 18, 405–409.

57. Maus, M.V. *et al.* (2002). *Ex vivo* expansion of polyclonal and antigen-specific cytotoxic T lymphocytes by artificial APCs expressing ligands for the T-cell receptor, CD28 and 4–1BB. *Nat Biotechnol* 20, 143–148.

58. Suhoski, M.M. *et al.* (2007). Engineering artificial antigen-presenting cells to express a diverse array of co-stimulatory molecules. *Mol Ther* 15, 981–988.

59. Golovina, T.N. *et al.* (2008). CD28 costimulation is essential for human T regulatory expansion and function. *J Immunol* 181, 2855–2868.

60. Zhang, H. *et al.* (2007). 4–1BB is superior to CD28 costimulation for generating CD8+ cytotoxic lymphocytes for adoptive immunotherapy. *J Immunol* 179, 4910–4918.

61. Liu, S. *et al.* (2007). IL-21 synergizes with IL-7 to augment expansion and anti-tumor function of cytotoxic T cells. *Int Immunol* 19, 1213–1221.

62. Huarte, E. *et al.* (2009). *Ex vivo* expansion of tumor specific lymphocytes with IL-15 and IL-21 for adoptive immunotherapy in melanoma. *Cancer Lett* 285, 80–88.

63. Zeng, R. *et al.* (2005). Synergy of IL-21 and IL-15 in regulating CD8+ T cell expansion and function. *J Exp Med* 201, 139–148.

64. Carroll, R.G. *et al.* (2008). Distinct effects of IL-18 on the engraftment and function of human effector CD8 T cells and regulatory T cells. *PLoS One* 3, e3289.

65. Emtage, P.C., Clarke, D., Gonzalo-Daganzo, R., and Junghans, R.P. (2003). Generating potent Th1/Tc1 T cell adoptive immunotherapy doses using human IL-12: Harnessing the immunomodulatory potential of IL-12 without the *in vivo*-associated toxicity. *J Immunother* 26, 97–106.

66. Brentjens, R.J. *et al.* (2003). Eradication of systemic B-cell tumors by genetically targeted human T lymphocytes co-stimulated by CD80 and interleukin-15. *Nat Med* 9, 279–286.

67. Naldini, L. *et al.* (1996). *In vivo* gene delivery and stable transduction of nondividing cells by a lentiviral vector. *Science* 272, 263–267.

68. Cavalieri, S. *et al.* (2003). Human T lymphocytes transduced by lentiviral vectors in the absence of TCR activation maintain an intact immune competence. *Blood* 102, 497–505.

69. Bai, Y. *et al.* (2003). Effective transduction and stable transgene expression in human blood cells by a third-generation lentiviral vector. *Gene Ther* 10, 1446–1457.

70. Chinnasamy, D. *et al.* (2000). Lentiviral-mediated gene transfer into human lymphocytes: Role of HIV-1 accessory proteins. *Blood* 96, 1309–1316.

71. Costello, E. *et al.* (2000). Gene transfer into stimulated and unstimulated T lymphocytes by HIV-1-derived lentiviral vectors. *Gene Ther* 7, 596–604.

72. Zhou, X. *et al.* (2003). Lentivirus-mediated gene transfer and expression in established human tumor antigen-specific cytotoxic T cells and primary unstimulated T cells. *Hum Gene Ther* 14, 1089–1105.

73. Gyobu, H. *et al.* (2004). Generation and targeting of human tumor-specific Tc1 and Th1 cells transduced with a lentivirus containing a chimeric immunoglobulin T-cell receptor. *Cancer Res* 64, 1490–1495.

74. Simmons, A., Whitehead, R.P., Kolokoltsov, A.A., and Davey, R.A. (2006). Use of recombinant lentivirus pseudotyped with vesicular stomatitis virus glycoprotein G for efficient generation of human anti-cancer chimeric T cells by transduction of human peripheral blood lymphocytes *in vitro*. *Virol J* 3, 8.

75. Yang, S. *et al.* (2008). Development of optimal bicistronic lentiviral vectors facilitates high-level TCR gene expression and robust tumor cell recognition. *Gene Ther* 15, 1411–1423.

76. Jones, S. *et al.* (2009). Lentiviral vector design for optimal T cell receptor gene expression in the transduction of peripheral blood lymphocytes and tumor-infiltrating lymphocytes. *Hum Gene Ther* 20, 630–640.

77. Yang, S., Rosenberg, S.A., and Morgan, R.A. (2008). Clinical-scale lentiviral vector transduction of PBL for TCR gene therapy and potential for expression in less-differentiated cells. *J Immunother* 31, 830–839.

78. Kalos, M. *et al.* (2011). T cells with chimeric antigen receptors have potent antitumor effects and can establish memory in patients with advanced leukemia. *Sci Transl Med* 3, 95ra73.

79. Porter, D.L., Levine, B.L., Kalos, M., Bagg, A., and June, C.H. (2011). Chimeric antigen receptor-modified T cells in chronic lymphoid leukemia. *N Engl J Med* 365, 725–733.

80. Frecha, C. *et al.* (2008). Stable transduction of quiescent T cells without induction of cycle progression by a novel lentiviral vector pseudotyped with measles virus glycoproteins. *Blood* 112, 4843–4852.

81. Buchholz, C.J., Muhlebach, M.D., and Cichutek, K. (2009). Lentiviral vectors with measles virus glycoproteins — dream team for gene transfer? *Trends Biotechnol* 27, 259–265.

82. Funke, S. *et al.* (2009). Pseudotyping lentiviral vectors with the wild-type measles virus glycoproteins improves titer and selectivity. *Gene Ther* 16, 700–705.

83. Levine, B.L. *et al.* (2006). Gene transfer in humans using a conditionally replicating lentiviral vector. *Proc Natl Acad Sci USA* 103, 17372–17377.

84. De Palma, M. *et al.* (2005). Promoter trapping reveals significant differences in integration site selection between MLV and HIV vectors in primary hematopoietic cells. *Blood* 105, 2307–2315.

85. Montini, E. *et al.* (2009). The genotoxic potential of retroviral vectors is strongly modulated by vector design and integration site selection in a mouse model of HSC gene therapy. *J Clin Invest* 119, 964–975.

86. Montini, E. *et al.* (2006). Hematopoietic stem cell gene transfer in a tumor-prone mouse model uncovers low genotoxicity of lentiviral vector integration. *Nat Biotechnol* 24, 687–696.

87. Williams, D.A. (2009). Gene therapy continues to mature and to face challenges. *Mol Ther* 17, 1305–1306.

88. Bastone, P. *et al.* (2007). Construction and characterization of efficient, stable and safe replication-deficient foamy virus vectors. *Gene Ther* 14, 613–620.

89. Mergia, A., Charl, S., Kolson, D.L., Goodenow, M.M., and Ciccarone, T. (2001). The efficiency of simian foamy virus vector type-1 (SFV-1) in nondividing cells and in human PBLs. *Virology* 280, 243–252.

90. Wickham, T.J. *et al.* (1997). Targeted adenovirus-mediated gene delivery to T cells via CD3. *J Virol* 71, 7663–7669.

91. Di Nicola, M. *et al.* (1999). Recombinant adenoviral vector-lipofectAMINE complex for gene transduction into human T lymphocytes. *Hum Gene Ther* 10, 1875–1884.

92. Chen, Z. *et al.* (2002). High-efficiency gene transfer to primary T lymphocytes by recombinant adenovirus vectors. *J Immunol Methods* 260, 79–89.

93. Schroers, R. *et al.* (2004). Gene transfer into human T lymphocytes and natural killer cells by Ad5/F35 chimeric adenoviral vectors. *Exp Hematol* 32, 536–546.

94. Moreau, A. *et al.* (2009). Efficient intrathymic gene transfer following *in situ* administration of a rAAV serotype 8 vector in mice and nonhuman primates. *Mol Ther* 17, 472–479.

95. Cooper, L.J. *et al.* (2006). Manufacturing of gene-modified cytotoxic T lymphocytes for autologous cellular therapy for lymphoma. *Cytotherapy* 8, 105–117.

96. Jensen, M.C. *et al.* (2000). Human T lymphocyte genetic modification with naked DNA. *Mol Ther* 1, 49–55.

97. Magg, T., Hartrampf, S., and Albert, M.H. (2009). Stable nonviral gene transfer into primary human T cells. *Hum Gene Ther* 20, 989–998.

98. Numbenjapon, T. *et al.* (2007). Antigen-independent and antigen-dependent methods to numerically expand CD19-specific CD8+ T cells. *Exp Hematol* 35, 1083–1090.

99. Park, J.R. *et al.* (2007). Adoptive transfer of chimeric antigen receptor redirected cytolytic T lymphocyte clones in patients with neuroblastoma. *Mol Ther* 15, 825–833.

100. Till, B.G. *et al.* (2008). Adoptive immunotherapy for indolent non-Hodgkin lymphoma and mantle cell lymphoma using genetically modified autologous CD20-specific T cells. *Blood* 112, 2261–2271.

101. Yaghoubi, S.S. *et al.* (2009). Noninvasive detection of therapeutic cytolytic T cells with 18F-FHBG PET in a patient with glioma. *Nat Clin Pract Oncol* 6, 53–58.

102. Huang, X., Wilber, A., McIvor, R.S., and Zhou, X. (2009). DNA transposons for modification of human primary T lymphocytes. *Methods Mol Biol* 506, 115–126.

103. Huang, X. *et al.* (2006). Stable gene transfer and expression in human primary T cells by the Sleeping Beauty transposon system. *Blood* 107, 483–491.

104. Singh, H. *et al.* (2008). Redirecting specificity of T-cell populations for CD19 using the Sleeping Beauty system. *Cancer Res* 68, 2961–2971.

105. Huang, X. *et al.* (2008). Sleeping Beauty transposon-mediated engineering of human primary T cells for therapy of CD19+ lymphoid malignancies. *Mol Ther* 16, 580–589.

106. Peng, P.D. *et al.* (2009). Efficient nonviral Sleeping Beauty transposon-based TCR gene transfer to peripheral blood lymphocytes confers antigen-specific antitumor reactivity. *Gene Ther* 16, 1042–1049.

107. Jin, Z. *et al.* (2011). The hyperactive Sleeping Beauty transposase SB100X improves the genetic modification of T cells to express a chimeric antigen receptor. *Gene Ther* 18, 849–856.

108. Singh, H. *et al.* (2011). Reprogramming CD19-specific T cells with IL-21 signaling can improve adoptive immunotherapy of B-lineage malignancies. *Cancer Res* 71, 3516–3527.

109. Nakazawa, Y. *et al.* (2009). Optimization of the PiggyBac transposon system for the sustained genetic modification of human T lymphocytes. *J Immunother.*

110. Nakazawa, Y. *et al.* (2011). PiggyBac-mediated cancer immunotherapy using EBV-specific cytotoxic T-cells expressing HER2-specific chimeric antigen receptor. *Mol Ther* 19, 2133–2143.

111. Birkholz, K. *et al.* (2009). Transfer of mRNA encoding recombinant immunoreceptors reprograms CD4+ and CD8+ T cells for use in the adoptive immunotherapy of cancer. *Gene Ther* 16, 596–604.

112. Lehner, M. *et al.* (2012). Redirecting T cells to Ewing's sarcoma family of tumors by a chimeric NKG2D receptor expressed by lentiviral transduction or mRNA transfection. *PLoS One* 7, e31210.

113. Mitchell, D.A. *et al.* (2008). Selective modification of antigen-specific T cells by RNA electroporation. *Hum Gene Ther* 19, 511–521.

114. Rabinovich, P.M. *et al.* (2006). Synthetic messenger RNA as a tool for gene therapy. *Hum Gene Ther* 17, 1027–1035.

115. Schaft, N. *et al.* (2006). A new way to generate cytolytic tumor-specific T cells: Electroporation of RNA coding for a T cell receptor into T lymphocytes. *Cancer Immunol Immunother* 55, 1132–1141.

116. Zhao, Y. *et al.* (2006). High-efficiency transfection of primary human and mouse T lymphocytes using RNA electroporation. *Mol Ther* 13, 151–159.

117. Zhao, Y. *et al.* (2010). Multiple injections of electroporated autologous T cells expressing a chimeric antigen receptor mediate regression of human disseminated tumor. *Cancer Res* 70, 9053–9061.

118. Barrett, D.M. *et al.* (2011). Treatment of advanced leukemia in mice with mRNA engineered T cells. *Hum Gene Ther* 22, 1575–1586.

119. Choi, Y. *et al.* (2010). A high throughput microelectroporation device to introduce a chimeric antigen receptor to redirect the specificity of human T cells. *Biomed Microdevices* 12, 855–863.

120. Hagani, A.B., Riviere, I., Tan, C., Krause, A., and Sadelain, M. (1999). Activation conditions determine susceptibility of murine primary T-lymphocytes to retroviral infection. *J Gene Med* 1, 341–351.

121. Pouw, N.M., Westerlaken, E.J., Willemsen, R.A., and Debets, R. (2007). Gene transfer of human TCR in primary murine T cells is improved by pseudo-typing with amphotropic and ecotropic envelopes. *J Gene Med* 9, 561–570.

122. Howland, L.J., Haynes, N.M., and Darcy, P.K. (2010). Generation of chimeric T-cell receptor transgenes and their efficient transfer in primary mouse T lymphocytes. *Methods Mol Biol* 651, 291–306.

123. Kerkar, S.P. *et al.* (2011). Genetic engineering of murine CD8+ and CD4+ T cells for preclinical adoptive immunotherapy studies. *J Immunother* 34, 343–352.

124. Lee, J., Sadelain, M., and Brentjens, R. (2009). Retroviral transduction of murine primary T lymphocytes. *Methods Mol Biol* 506, 83–96.

125. Klebanoff, C.A. *et al.* (2011). Determinants of successful CD8+ T-cell adoptive immunotherapy for large established tumors in mice. *Clin Cancer Res* 17, 5343–5352.

126. Pouw, N. *et al.* (2010). Combination of IL-21 and IL-15 enhances tumour-specific cytotoxicity and cytokine production of TCR-transduced primary T cells. *Cancer Immunol Immunother* 59, 921–931.

127. Gilham, D.E., Lie, A.L.M., Taylor, N., and Hawkins, R.E. (2010). Cytokine stimulation and the choice of promoter are critical factors for the efficient transduction of mouse T cells with HIV-1 vectors. *J Gene Med* 12, 129–136.

128. Movassagh, M. *et al.* (2000). Retrovirus-mediated gene transfer into T cells: 95% transduction efficiency without further *in vitro* selection. *Hum Gene Ther* 11, 1189–1200.

129. Kalos, M. (2012). Muscle CARs and TcRs: Turbo-charged technologies for the (T cell) masses. *Cancer Immunol Immunother* 61, 127–135.

130. Lipowska-Bhalla, G., Gilham, D.E., Hawkins, R.E., and Rothwell, D.G. (2012). Targeted immunotherapy of cancer with CAR T cells: Achievements and challenges. *Cancer Immunol Immunother.*

131. Hinrichs, C.S. *et al.* (2009). Adoptively transferred effector cells derived from naive rather than central memory CD8+ T cells mediate superior antitumor immunity. *Proc Natl Acad Sci USA* 106, 17469–17474.

132. Hinrichs, C.S. *et al.* (2011). Human effector CD8+ T cells derived from naive rather than memory subsets possess superior traits for adoptive immunotherapy. *Blood* 117, 808–814.

133. Ji, Y. *et al.* (2011). Repression of the DNA-binding inhibitor Id3 by Blimp-1 limits the formation of memory CD8+ T cells. *Nat Immunol* 12, 1230–1237.

134. Gattinoni, L. *et al.* (2011). A human memory T cell subset with stem cell-like properties. *Nat Med* 17, 1290–1297.

Chapter 3

TCR-Engineered T cells

Reno Debets[1] and Ton Schumacher[2]

[1]*Laboratory of Experimental Tumour Immunology,*
Department of Medical Oncology, Erasmus MC-Daniel den Hoed
Cancer Center, Rotterdam, The Netherlands
[2]*Division of Immunology, Netherlands Cancer Institute,*
Amsterdam, The Netherlands

3.1 Introduction

Adoptive T-cell therapy (ACT) with tumor-infiltrating lymphocytes (TILs) is currently the best treatment option for patients suffering from metastatic melanoma. ACT demonstrated objective clinical response rates of approximately 50% in case of TIL infusions, in conjunction with IL-2, were preceded by lymphoablative patient conditioning.[1] Equally interesting are the complete responses of 22% observed with TIL therapy, with ongoing complete regressions in the majority of patients with a follow-up of about 6 years or more.[2] Recently, a simpler and faster method was developed to generate "young" TILs that harbor characteristics associated with improved T-cell persistence and *in vivo* survival,[3,4] again showing objective response rates of about 50%. Next to TILs, adoptive transfer of tumor-specific T-cell clones generated from autologous peripheral T cells resulted in objective responses in metastatic melanoma patients.[5] In particular, peripheral T-cell clones that have been "educated" using artificial antigen-presenting cells yielded responses that were clinically long-lived and effective.[6] In contrast to ACT, the standard first-line treatment for melanoma (Dacarbazine[7]) or other types of immunotherapy tested in melanoma patients, such as vaccines and high dose IL-2, have demonstrated maximal response rates of about 10%.[8,9] Ipilimumab and MDX-1106 block cytotoxic T-lymphocyte antigen-4 (CTLA-4) and programmed death-1 receptor, respectively, and administration of these antibodies resulted in anti-tumor activities in metastatic melanoma and renal cell carcinoma (RCC) patients with response rates up to 28%.[10,11] Although results have encouraged further testing of dosing and combinations of these immune modulating reagents, currently responses are of a transient nature.

Despite its clinical successes, ACT has its limitations in availability and generation of therapeutic T cells for a larger group of patients. Genetic introduction of tumor-specific T-cell receptors (TCRs) into T cells, termed TCR gene therapy, can provide an alternative for ACT that is more widely applicable and can potentially be extended to several other types of cancer.[2,12,13] The principle of clinical TCR gene therapy is simple: one set of TCR genes can redirect many T cells toward the same peptide-major histocompatibility complex (MHC) target antigen that is expressed on diseased tissue. Already over two decades ago, it turned out to be technically feasible to confer a desired antigen specificity to T cells by means of transferring TCR genes.[14] Since then many efforts have been made to characterize and optimize the process of TCR gene transfer.[15–17] Generally, genetic engineering and adoptive transfer of T cells starts with the isolation of peripheral blood lymphocytes from a patient, followed by transduction with the TCR genes of interest and expansion to numbers that are needed clinically (i.e., $> 10^{10}$ T cells) within a period of up to 2 weeks prior to infusion in the patient. Up to now, six clinical trials using TCR-engineered T cells (in short TCR T cells) have been initiated and performed at the National Cancer Institute, National Institutes of Health, Bethesda, United States of America (see Table 3.1 for an overview of these trials with respect to TCRs used, cancers that were targeted and clinical observations), and additional trials are currently being initiated in other centers. The TCRs tested so far were all HLA-A2-restricted and directed against the melanoma differentiation antigens MART-1 or gp100, carcinoembryonic antigen (CEA), or the cancer testis antigen NY-ESO-1. Clinical responses have been

Table 3.1. Current overview of reported clinical TCR gene therapy trials.

Summary of six trials with TCR T cells, performed at NIH, Bethesda, NJ, US, with respect to antigen-specificity of TCR transgenes used (between brackets: code, affinity and origin of TCR); type of cancer that was treated; toxicities (between brackets: type of toxicity and number of patients who experienced toxicities out of total number of evaluable patients); T cell persistence as determined by flow cytometry with anti-TCR mAb using post-infusion peripheral blood mononuclear cells, and presented as the last timepoint by which a detectable flow cytometry result was obtained; objective responses as determined by RECIST criteria, and presented as percentage (between brackets: total number of evaluable patients); and references.

TCR specificity	Cancer	Toxicities	T-cell persistence	Objective response	Reference
MART-1/HLA-A2 (DMF4 TCR)	Metastatic melanoma	None (0/17)	12–13 months	12% (15)	18
MART-1/HLA-A2 (high affinity DMF5 TCR)	Metastatic melanoma	Severe melanocyte destruction in skin, eye, and ear (in some cases leading to uveitis and hearing loss) (9/36)	>1 months	30% (20)	
Gp100/HLA-A2 (murine TCR)	Metastatic melanoma		>1 months	19% (16)	19
CEA/HLA-A2 (murine TCR)	Metastatic colorectal carcinoma	Severe inflammation of colon (3/3)	>1 months	33% (3)	20
NY-ESO-1/HLA-A2 (high affinity IG4 TCR)	Metastatic melanoma	None (0/17)	>1 months	45% (11)	
	Metastatic synovial carcinoma			67% (6)	21

Abbreviations: CEA: carcinoembryonic antigen; Gp100: glycoprotein 100; HLA-A2: human leucocyte antigen A2; MART-1: melanoma antigen recognized by T cells 1; NY-ESO-1: New York esophageal cancer.

observed in patients with metastatic melanoma, colorectal, and synovial carcinoma.[18–21] These clinical trials have demonstrated feasibility, and initial responses were promising (see Table 3.1, and further discussed in Sec. 3.2). It should be noted, however, that therapy with TCR T cells resulted in severe inflammation of the skin, eyes, and ears (for MART-1 and gp100) or colon (for CEA),[19,20] likely due to low levels of target antigen expression in these organs. Chapter 4 provides an overview of the clinical application of T cells gene-modified with chimeric antigen receptors (CARs). From the clinical trials executed so far, one can deduce two challenges that require optimization to warrant further improvement of clinical TCR gene therapy, namely:

(a) self-reactivity of TCR T cells; and
(b) low functional avidity of TCR T cells.

These two challenges will both be discussed, together with potential solutions, in more detail in the Secs. 3.2 and 3.3. Section 3.4 discusses future perspectives of clinical TCR gene therapy.

3.2 Generation of TCR T Cells That are Not or Minimally Self-Reactive

3.2.1 *On-target toxicity*

Several ACT studies using expanded T-cell populations in mice[22–25] and men[26] as well as recent clinical receptor gene therapy studies[19,20,27] demonstrated the occurrence of toxicities of healthy tissues. For example, Lamers and colleagues treated RCC patients with T cells engineered with a CAR directed toward carbonic anhydrase IX (CAIX) and observed severe liver toxicity that was most likely due to expression of the CAIX antigen on the large bile ducts of the liver[27] (see also Chapter 4). In the study by Johnson and colleagues, skin, ear, and eye tissues were targeted by T cells engineered with a TCR directed against MART-1/HLA-A2 (MART-1/A2),[19] whereas in the study by Parkhurst and colleagues, colon tissue was targeted by T cells engineered with a TCR directed against CEA/A2.[20] Although such toxicities can often be suppressed and even reversed, these studies

demonstrated that T-cell therapy directed against tissue-differentia-
tion antigens cause (severe) side effects and suggest that therapy
should be directed against antigens with a (highly) tumor-restricted
expression pattern. Antigens that show tumor-specific expression are
those that belong to either mutated or shared antigens.[28] Mutated or
unique (neo)antigens, such as cyclin-dependent kinase-4 (CDK-4/m)
and BRAF(V600E), provide safe T-cell targets, yet the vast majority
of these antigens are expressed in individual tumors making a broad
clinical utility of these antigens at present difficult.[29] In contrast,
shared antigens, such as "cancer testis antigens" (CTA), are expressed
in many tumors, often by promoter demethylation, but silenced in
normal adult cells except immune-privileged male germline cells and
thymic medullary epithelial cells.[30,31] Preclinical and clinical studies
have shown the immunogenic nature of these antigens (eloquently
reviewed in Ref. 32). Already in the early 1990s, van der Bruggen and
colleagues identified MAGE-A1 as the first immunogenic tumor antigen, which triggered cytotoxic T-cell responses in a cancer patient.[33]
So far, there are over 110 identified combinations of CTA peptides
and HLA molecules recognized by T cells, of which approximately 30
peptides are encoded on the X-chromosome. In particular, these latter CTAs have been recognized for their active contribution to the
development of malignancies.[34-36] In addition, *in vitro* studies have
shown that gene transfer of TCRs directed against MAGE-A1/A1,
MAGE-C2/A2, NY-ESO-1/A2 as well as MAGE-A3/DP4 and
NY-ESO-1/DP4 can result in effective T-cell responses against tumor
cells expressing these CTAs.[37-39] Notably, a recent clinical trial with
T cells gene-engineered to express a TCR directed against NY-ESO-1
provided evidence for objective responses in 5 out of 11 patients with
metastatic melanoma and 4 out of 6 patients with synovial carcinoma,
with no signs of toxicity (see Table 3.1). Collectively, these studies
advocate the therapeutic testing of T cells that target neoantigens,
such as those expressed due to epigenetic alterations in tumors.

Other criteria that can be taken into account when defining a
T-cell target antigen, in more detail discussed in Chapter 1, are the
antigen's ability to be cross-presented on tumor stroma,[40,41]) and
whether or not an antigen is expressed by tumor initiating cells.[42]

3.2.2 *TCR mis-pairing and potential off-target toxicity*

TCR mis-pairing is a recognized phenomenon in the field of TCR gene therapy, which defines the incorrect pairing between TCRα or β transgenes and endogenous TCRβ or α chains, respectively. Although there is currently no clinical evidence for TCR mis-pairing-induced autoreactivity, the generation of autoreactive TCRs upon TCR mis-pairing cannot be excluded. In fact, mouse T cells expressing mis-paired TCRs and expanded under high IL-2 conditions (similar to the current clinical setting) were demonstrated to induce graft-versus-host disease (GvHD) in a preclinical model.[43] In addition, human T cells expressing mis-paired TCRs demonstrated neo-reactivity *in vitro*.[44] Strategies that increase preferential TCR pairing and counteract TCR mis-pairing are anticipated to circumvent the generation of unknown TCR specificities and the development of potential off-target autoimmune reactivity. Various strategies have been developed that address TCR mis-pairing, such as (i) genetic modification of TCR transgenes; (ii) disruption of endogenous TCRα and β genes; and (iii) TCR gene transfer into defined T-cell populations.

3.2.3 *Genetically modified TCRs*

In this chapter, we have defined five classes of genetic strategies to modify TCR transgenes and control their pairing properties, which have been schematically depicted in Figure 3.1 (see Ref. 45 for a review).

3.2.3.1 *Murinization of TCR chains*

In murinized TCRs, the human TCR constant (TCR-C)α and TCR-Cβ domains have been replaced by the corresponding murine TCR-C domains. Although human and murine TCR-C domains show a high degree of homology, it is believed that small differences affect the stability of TCR/CD3 interactions and hence TCR surface expression levels. Evidence for a competitive advantage of murinized TCR for TCR/CD3 surface expression comes from the observation

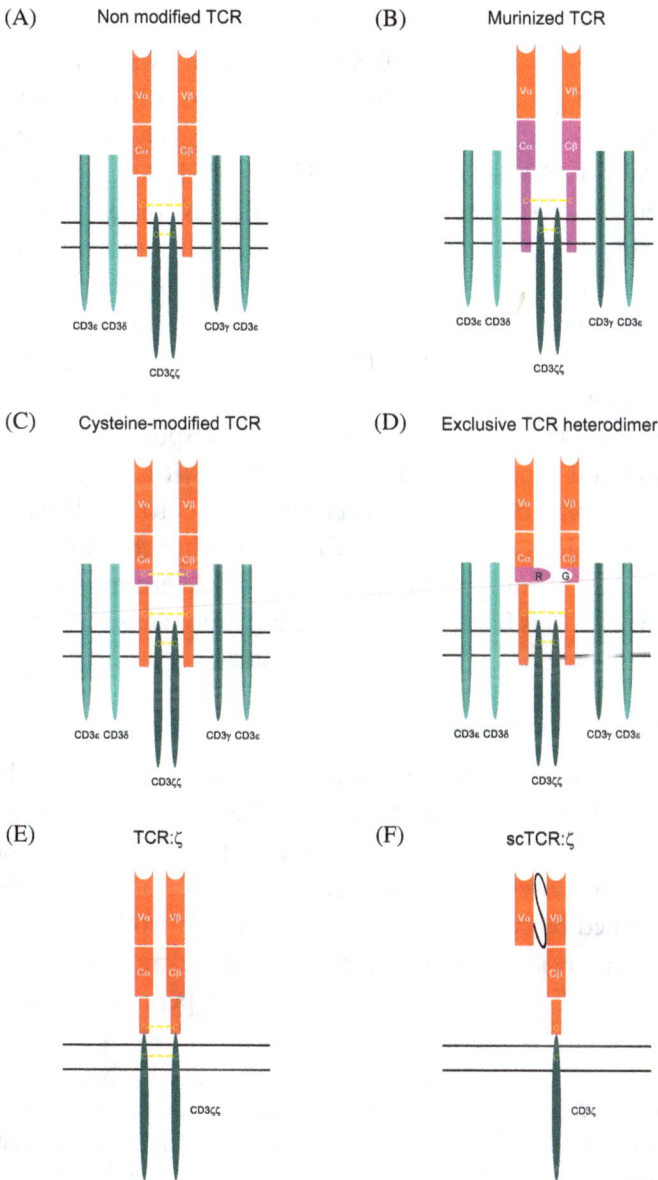

Figure 3.1. Illustration of various genetically modified TCR constructs to address TCR mis-pairing.

The simple transfer of non-modified TCR (A) results in TCR mis-pairing between the endogenous and introduced TCR chains. TCRs that are gene-modified to reduce that murine TCR-C domains bind human CD3ζ more strongly than human TCR-C domains.[46] TCR murinization as a means to address TCR mis-pairing has initially been investigated using the MDM2/A2 and WT1/A2 specificities[47,48] and more extensively using the p53/A2 and MART-1/A2 specificities.[46] These studies generally demonstrated enhanced surface expression and pMHC binding of murinized TCRs and, in some cases, enhanced T-cell functional responses toward tumor cells. Further efforts aimed at minimizing the number of murine amino acids with the intent to reduce the potential risk for immunogenicity of the TCR transgenes. These studies identified a limited set of amino acids (a Lys in mouse Cβ, and a stretch of Ser, Asp, Val, and Pro in mouse Cα) that are responsible for improved expression and function of murine TCRs.[49,50] It is noteworthy, however, that despite its positive effects with respect to T-cell avidity, TCR murinization does not address TCR mis-pairing to a full extent, and the level of TCR mis-pairing may in fact be determined by sequences of the endogenous TCR variable (TCR-V) domains.[46]

3.2.3.2 Cysteine-modified TCR

Introduction of cysteine amino acids at structurally favorable positions allows formation of an additional disulfide bridge and promotes correct pairing between the TCRα and β chains.[51] Site-directed mutations of Cα Thre48Cys and Cβ Ser57Cys resulted in WT1/A2 TCR$\alpha\beta$ linked by two inter-chain bonds (i.e., introduced plus endogenous cysteines).[52] Introduction of cysteine-modified TCR$\alpha\beta$ in human CD8 T cells resulted in increased peptide–MHC binding

---

Figure 3.1. (*Continued*) their ability to mis-pair include: TCRs with murine TCR-C domains (B), TCRs with extra disufide bond (C), TCR:ζ (D), and single-chain TCRs (E). See text in Sec. 3.2 for details. Explanation of color-coding in cartoons: TCR$\alpha\beta$ heterodimer with variable (V) and constant (C) domains in red; endogenous cysteine bonds (in A) and additional cysteine bond (in C) in yellow; CD3$\gamma\delta\varepsilon\zeta$ molecules in green; murine constant domains and ectopic mutations (in C and D) in purple; linker between variable domains as a black line (in F). (Figure is modified, with permission, from Govers, *Trends Mol Med*, 2010.)

compared to non-modified (wild type) TCR$\alpha\beta$, which corresponded to enhanced peptide-specific T-cell cytotoxicity and IFNγ production.[48,52] Cysteine-modified TCRs show enhanced TCR pairing, but their ability to mis-pair was at best modestly decreased when compared to non-modified TCRs.[48,52] When combining murinization and cysteine modification, expression and functional data look promising,[53] yet to date minimal data regarding TCR mis-pairing of such dual-modified TCRs are available.

3.2.3.3 Exclusive TCR heterodimer

To generate an exclusive TCR heterodimer, steric and electrostatic forces have been exploited to inhibit TCR mis-pairing and at the same time facilitate correct pairing between TCRα and β chains. The crystal structures of murine 2C/H-2K^b and human Tax/A2 TCRs identified Ser85Arg for TCR-Cα and Arg88Gly for TCR-Cβ as mutations to yield the required changes in electrostatic charges. These mutations were expected to generate a reciprocal "knob-into-hole" configuration, and to minimally distort secondary and tertiary structures.[54] Experimentally, mutated TCR chains did show a reduced ability to mis-pair, which was accomplished at the expense of correct TCR pairing, tetramer binding, and antigen-specific cytolysis, but not IFN-γ secretion, in primary human T cells.[47]

3.2.3.4 TCR coupled to CD3ζ

TCRα and β chains both fused to a complete human CD3ζ molecule (abbreviated as TCR:ζ) endow TCR chains with CD3ζ-mediated dimerization between the two TCR chains. Already in 2000, it was reported that transduction of primary human T cells with either single TCRα:ζ or TCRβ:ζ does not result in cell surface expression, suggesting that these TCR chains lack the ability to mis-pair with endogenous TCR chains.[37] In a subsequent study, flow cytometric measurement of fluorescent resonance energy transfer (FRET) between corresponding as well as non-corresponding TCRα and TCRβ chains demonstrated high preferential pairing for TCR:ζ in the absence of TCR mis-pairing.[55] Immune precipitation studies revealed that TCR:ζ did not associate with CD3ε, CD3γ, CD3δ, and only marginally with CD3ζ. Consequently, TCR:ζ does not compete with endogenous TCRαβ for available CD3 molecules, and shows improved cell surface expression.[55] Regarding functional assays, such as NFAT activation, cytotoxicity, and cytokine production, TCR:ζ performed equally well than non-modified TCR.[37,55,56]

3.2.3.5 *Single-chain TCR*

A single-chain (sc)TCR combines the variable domains of TCRα and β into one chain. Generally, a TCR-Vα domain is attached to a TCR-Vβ domain, interspersed by a linker sequence, and followed by a TCR-Cβ domain which is partially replaced by CD3ζ (or the CD3-like Fc(ε)RIγ molecule) to provide downstream signaling and T-cell activation[37,57] (abbreviated as scTCR:ζ). A scTCR:ζ based on a parental MAGE-A1/A1-specific TCR, derived from the CTL 82/30,[33] led to an increased surface expression but a reduced peptide–MHC binding as well as antigen-specific T-cell functions when compared to two-chain TCR:ζ.[37] Other researchers have built on this concept, and showed that scTCRs not molecularly linked to CD3ζ do not prevent mis-pairing with endogenous TCR chains,[58] and developed additional strategies to enhance the functional avidity of scTCR T cells. In one such strategy, a high-affinity TCR has been combined with enhanced signaling cassettes that contain CD3ζ, CD28 and LCK, and resulted in scTCRs that function equally well compared to non-modified TCR in primary mouse T cells.[58] In another strategy, scTCR (with murine TCR-Cβ) was co-expressed with the murine TCR-Cα domain and, in particular when both chains were cysteine-modified, provided human T cells with an ability to respond toward tumor cells *in vitro* and *in vivo* that was similar, and in some cases slightly better, than T cells expressing non-modified TCR.[59]

3.2.4 *Editing T-cell specificity*

The above-mentioned gene-modified TCRs, although in some cases showing promising results, leave the endogenous TCR chains untouched. Disruption of endogenous TCR chains may represent a cleaner system that would formally rule out the occurrence of TCR mis-pairing. To this end, two approaches have been experimentally tested. One approach relied on small interfering (si)RNAs to specifically downregulate endogenous TCRα and β chains in combination with siRNA-resistant TCR transgenes. This approach resulted in enhanced functional avidity of T cells gene-engineered with human TCRs directed against either MAGE-A4 or WT1 *in vitro*, [60] and mouse TCR against LCMV gp33 *in vitro* and *in vivo* (Prof. Wolfgang Uckert, MDC, Berlin, personal communication). The silencing of endogenous TCR genes, however, was not complete as evidenced by flow cytometry and off-target activity *in vivo*. Recently, another approach has been elegantly tested by Chiara Bonini and colleagues, who deleted endogenous TCRα and β genes by zinc-finger nucleases, enriched TCR-deleted T cells and edited these T cells with a new WT-1 TCR.[61] The newly generated T cells showed an enhanced functional avidity when compared to non-edited TCR T cells, and lacked any off-target activity *in vivo*.[61] The complete editing of T cells is currently challenged by a multi-step T-cell processing method that includes sequential T-cell stimulations, zinc-finger treatments, and T-cell sorts and covers a period of about 5 weeks. The rationale, however, successfully yielded effective and safe T cells, which, together with a more simplified method, holds great promise for T-cell therapy.

3.2.5 *T-cell populations with restricted or no TCR usage*

In addition to the use of genetically modified TCRs or TCR-edited T cells, a third method to address TCR mis-pairing may be the use of defined host T-cell populations for TCR gene transfer.[62,63] TCR gene transfer into restricted T-cell populations is anticipated to reduce the risks of autoimmune reactivities for two reasons: (i) recipient T cells are non-self reactive and (ii) chances that new TCR reactivities are generated as a consequence of TCR mis-pairing is reduced or even absent. However, it should be noted that the choice of defined T-cell populations might affect the anti-tumor efficacy of TCR T cells. For example, TCR$\alpha\beta$-engineered $\gamma\delta$T cells showed no evidence of TCR mis-pairing, yet demonstrated compromised anti-tumor efficacy in an OVA mouse model when compared to TCR$\alpha\beta$-engineered $\alpha\beta$T cells.[63]

3.2.6 *Removal of adoptively transferred T cells*

Despite the availability of various means to address TCR mis-pairing, one may choose for a (additional) strategy to eliminate potentially autoreactive T cells. Elimination of T cells can be divided according to the mechanism of removal: (i) drug-induced T-cell suicide and (ii) tag-mediated T-cell killing. Drug-induced T-cell suicide is based on genetic introduction of a suicide gene that can be activated by addition of a specific substrate or ligand, such as ganciclovir for the herpes simplex virus-thymidine kinase suicide gene[64] or FK506 dimers for an inducible caspase 9 (iCasp9$_M$).[65,66] Tag-mediated T-cell killing is based on complement-mediated clearance following treatment with antibodies directed at a CD20[67] or myc-epitope that is genetically incorporated into a TCR chain.[68] The mentioned strategies show high potential in eradicating transduced T cells but may pose drawbacks, such as the induction of immune responses against suicide genes or tags, the development of gene-loss variants, and the high costs often involved in the good manufacturing practice (GMP)-preparation of necessary antibodies.

3.3 Generation of TCR T Cells with Enhanced Functional Avidity

T-cell responsiveness of TCR T cells is considered less when compared to non-modified T cells and can be improved via various strategies, such as the use of (i) optimal vectors and transgene cassettes; (ii) genetically modified TCRs; and (iii) multiple TCRs.

3.3.1 *Vectors and transgene cassettes*

Procedures for retroviral gene transfer have been optimized for both murine T cells[69] and human T cells, the latter at GMP level.[70] Here, we will focus on advances that have been made with respect to the make-up of the vector backbone and transgene cassette. Most TCR gene transfer studies published to date have made use of gammaretroviral vectors for transgene delivery, and recently it has also been shown feasible to transduce TCR transgenes using a lentiviral system.[71] Clinically, there is experience with TCR vectors that contain murine stem cell virus (MSCV), long terminal repeats (LTRs) and optimized splice and start codons.[18-21] See Chapter 2 for more details on gene transfer methodology and the use of other virus vehicles. There is now accumulated preclinical evidence that transduction efficiencies differ substantially between different vectors, which is at least in part attributed to the viral origins of LTRs and different splice and start sequences. In this respect, it of interest to mention that the pMP71 vector, which has a myeloproliferative sarcoma virus (MPSV) LTR and optimal 5′ sequences, demonstrated highly improved TCR transduction efficiencies.[72,73] In clinical vectors, TCRα and β genes are encoded by a single construct and separated either by internal ribosomal entry site (IRES) or 2A sequences.[19] IRES sequences are used to achieve promoter-independent protein translation of the second gene, which in some cases is less relative to the first gene product.[74] 2A sequences lead to single mRNA molecules that are processed into two individual proteins by endoproteases, and generally result in equal levels of surface expression of both gene products.[75,76] The orientation of the TCRα and β chains within a vector may be of importance. Recent work suggests that placing the TCRβ chain in front of the TCRα chain, especially when separated by a 2A sequence, renders optimal functional TCR expression levels for most TCRs tested.[77]

An important safety issue concerning retroviral gene transfer is linked to insertional mutagenesis as evidenced by reports of leukemia as a result of treating X-SCID patients with CD34$^+$ progenitor cells transduced with common γ-chain.[78] It is of note that mature T cells in contrast to hematopoietic progenitor cells are resistant to oncogenic transformation.[79] Also clinically, there are > 500 patient-years of follow-up upon treatment with high numbers of retrovirally modified mature T cells demonstrating that decay half-lives of modified T cells exceeded 16 years with no evidence of genotoxicity and virus integration sites near genes implicated in tumorigenesis.[80]

3.3.2 Genetically modified TCR formats

TCR molecules can be genetically engineered to enhance T-cell surface expression and function. To this end, multiple strategies have been reported which we have classified as strategies that enhance (i) surface expression; (ii) ligand-binding affinity, and/or (iii) TCR's signaling potency.

3.3.2.1 Codon optimization

Codon usage (i.e., amino acids are encoded by multiple codons) and the presence of secondary structures within the coding region of genes have been reported to affect gene expression and such factors may also be taken into account when designing synthetic genes for (TCR) gene transfer.[81] Recent data show that codon optimization of TCR genes has a beneficial effect on surface expression and *in vitro* and *in vivo* function of TCR T cells.[82,83]

3.3.2.2 Mutations in TCR transmembrane and constant domains

Engineering TCR transmembrane and constant domains affects the functional avidity of TCR T cells. For instance, incorporation of the hydrophobic residues Leu, Val, Leu at evolutionary permissive positions in the transmembrane region of TCRα improved the TCR's stability and the functional avidity of T cells.[84] In addition, decreased N-glycosylation of the TCR has been reported to enhance binding of peptide–MHC and to result in improved functional T-cell avidity as measured by cytokine release and lytic activity.[85] Deglycosylation of the TCR-C domain is anticipated to decrease the T-cell activation threshold, possibly as a consequence of improved membrane movement or multimerization of the TCR. This glycosylation technique was found to be effective for multiple TCRs without evidence for self-reactivity.

3.3.2.3 *Strategies that enhance the TCR's ligand-binding affinity*

Tumor-specific TCRs often are at the lower affinity end of the natural TCR repertoire. *In vitro* display of TCR libraries has been successfully applied toward the selection of affinity-enhanced TCRs. In a recent phage display study to select high affinity HIV-1-specific TCR, the ligand binding affinity could be improved from the nM to pM range.[86] In addition to phage display technology, mutated TCR variable domains have also been screened via RNA-based transfections, which yielded affinity-improved MART-1/A2 and NY-ESO-1/A2 TCRs following single or dual amino acid substitutions in the complementarity determining region (CDR).[87] Along this example, an affinity-enhanced NY-ESO/A2 TCR (derived from T cell clone 1G4) has been successfully used in clinical TCR gene therapy.[21] Another angle to obtain TCRs with high affinities is to start from a pool of non-tolerant TCRs. In example, the use of MHC-mismatched CTL *in vitro* systems allows the isolation of tumor-specific TCR genes from healthy donor-derived material and circumvents self-tolerance.[88] Immunization of HLA-transgenic mice with tumor antigens would also circumvent self-tolerance. Both the gp100 and CEA-specific TCRs used in clinical trials were isolated from HLA-A2 transgenic mice.[19,20] Not completely unexpected, patients in trials using murine TCRs developed antibodies against non-human TCRs.[89] Recently, the group of Prof. Thomas Blankenstein (Max-Delbrück-Center for Molecular Medicine, Berlin, Germany) put great effort into the generation of a mouse in which the complete human TCR loci were introduced while the murine TCR loci were inactivated.[90] This mouse model can be used to isolate complete human TCRs from a non-tolerant repertoire directed against those human peptides that differ between mice and humans. An advantage of this model would be that resulting TCRs are not immunogenic in humans. Currently, these mice express HLA-A2 molecules but any desired human MHC molecule can be introduced making this model a potentially valuable source of TCRs that can be used in TCR gene therapy.

3.3.2.4 *Introduction of additional CD3 components*

In TCR T cells, the introduced and endogenous TCRs compete with available CD3 chains for surface expression. This competition for CD3 chains has been addressed by enhancing the levels of CD3 molecules in T cells (also see the above-mentioned TCR:ζ, a CD3-independent TCR, representing another strategy that addresses competition for CD3 chains). Indeed, co-transfer of CD3 and TCR genes into primary murine T cells resulted in improved levels of TCR expression and allowed T cells to respond to lower concentrations of antigen.[91] Also, T cells co-expressing TCR and additional CD3 molecules resulted in faster tumor infiltration, tumor elimination, and an enhanced memory T-cell response.

3.3.2.5 *Strategies that enhance the TCR's signaling potency*

In extension to the genetic introduction of CD3ζ (or CD3-like domains) into receptors, as discussed for TCR:ζ, other "building blocks" constituting transmembrane and/or intracellular domains of accessory molecules, co-stimulatory molecules, and kinases have been analyzed for their effect on surface expression and function of various receptors. Receptors, including scTCR, that contain the transmembrane domain of CD3ζ followed by the intracellular domain of CD28 and the co-receptor molecule Lck (i.e., scTCR:ζ -28-Lck) constitute a promising format with respect to T-cell activation and peptide-specific T-cell functions.[58,92,93] Unless serious efforts are put into place to optimize scTCRs, it has generally been observed that these receptors were less responsive than non-modified TCR (see Sec. 3.2). The CD28 co-stimulatory domain has also been incorporated in TCR:ζ, which resulted in enhanced tumor-specific IFNγ production of transduced T cells.[94,95] In addition to CD28, the incorporation of other TNF receptor super family members, such as CD134 and CD137, into TCR molecules may further optimize T-cell functions as suggested by the preclinical and clinical testing of CARs that incorporate such co-stimulatory molecules.[96,97]

3.3.3 *Simultaneous use of multiple TCRαβ transgenes*

Limited (i.e., single) antigen specificity is inherent to gene transfer of a single TCR transgenes (whereas TILs mostly cover a few antigen specificities). Restricted antigen reactivity may have contributed to compromised clinical responses following adoptive transfer of TCR T cells when compared to TILs. In addition, targeting tumors with a T-cell population specific for a single antigen may lead to the selective outgrowth of antigen-negative tumor variants.[5,82] To reduce the risk of immune escape, one could opt to create multiple T-cell pools with different specificities using a different TCRαβ per T-cell pool. Successful elimination of cancers may require the cooperation of CD4 and CD8 T cells not only during the induction phase but also during the effector phase in the tumor microenvironment.[98] Hence, simultaneous use of both MHC class I- and II-restricted TCRs in CD8 and CD4 T cells, respectively, directed at the same antigen or different antigens may further enhance the efficacy of ACT.

3.3.4 *Other strategies*

In addition to vector design, gene-modified TCRs, and the use of multiple TCRs, T-cell responsiveness can be enhanced by strategies that have primarily been designed to enhance the peripheral persistence of T cells. Peripheral persistence of TCR T cells is decreased when compared to TILs following adoptive transfer[18,19,99](see Table 3.1) and can be improved via various strategies, such as the use of (i) less differentiated CD8 T cells; (ii) CD4 T-cell help; and (iii) dual-specific T cells. These strategies are shortly described below, but more detailed information on T-cell persistence can be found in Chapter 1.

3.3.4.1 *Less differentiated CD8 T cells*

T-cell persistence is reported to be inversely associated with differentiation state and replicative history of transferred T cells.[100] One way to obtain less differentiated T cells is to expose T cells to common γ cytokines other than IL-2 prior to adoptive T-cell transfer. For example, treatments with either IL-15+IL-21 or IL-7+IL-15 have been shown to generate gene-engineered T cells with a less differentiated CD8 T cell phenotype (i.e., central memory phenotype), prolonged peripheral persistence, and potent antigen reactivity.[101,102] Alternatively, one can use less differentiated T-cell populations as recipient cells for gene transfer.[103,104] Along this line, a recently identified population of "stem-cell memory" CD8 T cells, expressing high levels of CD95, IL2Rβ, and demonstrating increased proliferative potential and ability to mediate anti-tumor responses, may represent a subset of T cells, which may best fulfill these criteria.[105]

3.3.4.2 *CD4 T-cell help*

Administration of CD4 T-helper cells concurrently with CD8 T cells has been shown to prevent exhaustion of infused CD8 T cells[106,107] and to result in effective anti-tumor T-cell responses.[108] Notably, CD4 T cells producing Th1-type cytokines, but not those producing Th2-type cytokines, have been reported to eradicate tumors in T-cell transfer studies.[109,110] CD4 T cells can be functionally endowed with MHC I-restricted TCR$\alpha\beta$ via gene transfer,[111,112] and genetic co-introduction of CD8α may skew TCR T cells toward an antigen-specific Th1-type T-cell response.[113] However, *in vivo* expansion of antigen-challenged IFNγ^{low} CD4 T cells may be required prior to development of IFNγ^{high} CD4 T cells and synergy with CD8 T cells to build a potent anti-tumor response.[114] Furthermore, adoptive transfer of Th17-polarized CD4 T cells effectively mediates rejection of TRP-1 positive tumors in a TCR-transgenic mouse model.[115] CD4 Th17 cells appear to be long-lived and their molecular signature resembles that of "stem-cell memory" CD8 T cells.[116] Interestingly, the anti-tumor activity of CD4 Th17 cells depends on its (incomplete) differentiation and conversion into Th1 cells, resulting in a co-existence of Th17 and Th1 cells, and it is this multi-potency that provides CD4 T-cell subset its therapeutic potential.

3.3.4.3 *Dual-specific T cells*

Already in 2002, a strategy was developed that makes use of highly avid monoclonal CD8 T cells as recipient cells for receptor gene transfer which resulted in enhanced peripheral T-cell persistence and sustained immune responses against tumors.[117] The generation of dual-specific T cells is expected to result in ongoing T-cell stimulation via the endogenous (often anti-virus) TCR and prevention of tumor-induced T-cell anergy.[117,118] In fact, treatment of neuroblastoma patients with EBV-specific CTL expressing a receptor directed against disialoganglioside GD2 showed enhanced T-cell survival and tumor regression in half of the patients treated[118,119] (see also Chapter 4).

Strategies that create T cells that are non-self-reactive yet highly functionally avid toward the cognate antigen, as discussed in Secs. 3.2 and 3.3, are considered necessary to further improve the safety and efficacy of clinical TCR gene therapy. These strategies are summarized in Table 3.2.

3.4 Next Steps in TCR Gene Therapy

Several clinical trials using TCR-T cells have been completed (see Table 3.1), and a significant number of trials are planned. Although remarkable successes have been observed, it still remains to be formally proven that TCR gene therapy delivers improvements in both progression-free and overall survival when compared to standard treatment of care. To address this issue, randomized trials with larger numbers of patients at multiple participating centers need to be performed. Another challenge is the large diversity of current trials making it potentially difficult to identify single parameters important to the clinical success of TCR gene therapy. To address this second issue, multiple single parameter and small-scale trials need to be performed to accelerate decision making and facilitate the design of an optimal (combined) protocol, the latter to be tested in large-scale trials.[120] In addition to the above-mentioned parameters (Secs. 3.2 and 3.3, and summarized in Table 3.2), other parameters considered important for the future development of clinical TCR gene therapy are related to TCR selection, T-cell processing, and clinical design (see also Chapters 2 and 4–6).

Table 3.2. Gene-engineering to further improve TCR gene therapy.
An overview of gene-engineering strategies that can be employed to improve the
efficacy and safety of TCR gene therapy, as discussed in this chapter. Note that the
presented list of strategies is not complete, and other strategies such as those aimed
at antigen choice (and hence choice of TCR), T-cell subsets and differentiation state,
and combination therapies are described in more detail in Secs. 3.2 and 3.3.

Toxicity	**On-target toxicity** Choose target antigens that demonstrate a highly or unique tumor-restricted expression. **Off-target toxicity** **(note: mentioned strategies also affect functional T-cell avidity)** *Modify TCR format to prevent/minimize TCR mis-pairing:* * Murinized TCR * Cysteine-modified TCR * Exclusive TCR heterodimer * TCR:ζ * Single-chain TCR (or variants thereof). *Editing of T-cell specificity:* * siRNA * Zinc-fingers nucleases. *Defined T-cell populations to prevent/minimize TCR mis-pairing:* * Cytotoxic T-cell clone with known endogenous TCR * γδ T cells.
Functional T-cell avidity	**Vector/Transgene cassette** Choose suitable vector for high-level gene expression, combined with optimal TCR transgene cassette. **Enhance expression/function of TCR** * Codon optimization * Mutate TCR transmembrane or constant domains * Introduce additional CD3 components * Enhance TCR's signaling potency. **Enhance ligand-binding affinity of TCR** * Mutate TCR variable domains * Isolate TCRs from non-tolerant TCR repertoires. **Simultaneous use of multiple TCRs** Multiple T-cell pools with a different TCR per T cell pool.

3.4.1 Selection of TCR genes for clinical application

Tumor-specific TCRs for use in immune therapy can be obtained through various methods that can be roughly divided into (i) *in vitro* generation of tumor-specific T cells (as a source of TCRs), and (ii) *in vitro* selection of TCR genes. With respect to *in vitro* generation of tumor-specific T cells, dominant TCRs have been isolated from T-cell clones generated from polyclonal T-cell populations, such as TILs. With respect to *in vitro* selection of TCR genes, TCRs have been isolated from single cells that were sorted following labeling of peripheral blood T cells with peptide–MHC complexes.[87,121,122] In example, MART-1[18,19,99] and NY-ESO-1 TCRs[21] have been obtained and characterized according to these two methods and used for clinical TCR gene therapy. To facilitate a deliberate choice for tumor target antigens and corresponding TCRs for future clinical use, new techniques are required that would allow an efficient and high-throughput detection and isolation of tumor-specific T cells within therapeutic T-cell populations. To this end, elegant techniques have been developed by the group of Prof. Ton Schumacher (National Cancer Institute, Amsterdam) that are based on UV-mediated peptide–MHC exchange,[123] combinatorial encoding of peptide–MHC complexes,[124] and capture and analysis of TCR genes.

3.4.2 T-cell processing and clinical design

Production of TCR T cells is expensive and can only be performed in a relatively restricted number of specialized centers, with a logistic and financial need to improve and simplify the manufacturing process. Major issues for patient therapy remain the preconditioning, T-cell dosing, and post-transfer cytokine support. Generating more effective, more fitter T cells means that patients may require less intensity preconditioning, T-cell dosing, and cytokine support.[125] National and international grant support in the specific area of T-cell processing and clinical design, areas that are often not easily covered by regular calls, would benefit the field of TCR gene therapy.

At the moment of writing this chapter, two multi-center trials will be EU-sponsored (EU Framework 7 project, acronym ATTACK) and executed shortly, both with T cells gene-engineered to express an NY-ESO-1 TCR (see also Chapters 5 and 6). One trial will be a standard Phase II trial in esophago-gastric cancer patients who have disease after standard therapy. Another trial is a randomized Phase II study in patients with advanced melanoma to assess whether sorting for CD62L-positive T cells, T-cell activation with anti-CD3 and CD28 mAbs, and T-cell expansion with IL-7 and IL-15 will produce gene-modified T cells that show a superior capacity to repopulate in patients. These trials, and other trials like these, are expected to provide answers to some of the above-mentioned questions and will help this promising field further.

Take home message: To ensure further clinical development of TCR gene therapy, it is necessary to choose safe T-cell target antigens, and implement (combinations of) strategies that enhance the correct pairing of TCR transgenes as well as the functional avidity and persistence of T cells.

References

1. Dudley, M.E., Yang, J.C., Sherry, R., Hughes, M.S., Royal, R., Kammula, U., Robbins, P.F., Huang, J., Citrin, D.E., Leitman, S.F., Wunderlich, J., Restifo, N.P., Thomasian, A., Downey, S.G., Smith, F.O., Klapper, J., Morton, K., Laurencot, C., White, D.E., and Rosenberg, S.A. (2008). Adoptive cell therapy for patients with metastatic melanoma: Evaluation of intensive myeloablative chemoradiation preparative regimens. *J Clin Oncol* 26, 5233–5239.
2. Restifo, N.P., Dudley, M.E., and Rosenberg, S.A. Adoptive immunotherapy for cancer: Harnessing the T cell response. *Nat Rev Immunol* 12, 269–281.
3. Besser, M.J., Shapira-Frommer, R., Treves, A.J., Zippel, D., Itzhaki, O., Hershkovitz, L., Levy, D., Kubi, A., Hovav, E., Chermoshniuk, N., Shalmon, B., Hardan, I., Catane, R., Markel, G., Apter, S., Ben-Nun, A., Kuchuk, I., Shimoni, A., Nagler, A., and Schachter, J. Clinical responses in a phase II study using adoptive transfer of short-term cultured tumor infiltration lymphocytes in metastatic melanoma patients. *Clin Cancer Res* 16, 2646–2655.
4. Dudley, M.E., Gross, C.A., Langhan, M.M., Garcia, M.R., Sherry, R.M., Yang, J.C., Phan, G.Q., Kammula, U.S., Hughes, M.S., Citrin, D.E., Restifo, N.P., Wunderlich, J.R., Prieto, P.A., Hong, J.J., Langan, R.C., Zlott, D.A., Morton, K.E., White, D.E., Laurencot, C.M., and Rosenberg, S.A. CD8+ enriched "young" tumor infiltrating lymphocytes can mediate regression of metastatic melanoma. *Clin Cancer Res* 16, 6122–6131.

5. Yee, C., Thompson, J.A., Byrd, D., Riddell, S.R., Roche, P., Celis, E., and Greenberg, P.D. (2002). Adoptive T cell therapy using antigen-specific CD8+ T cell clones for the treatment of patients with metastatic melanoma: In vivo persistence, migration, and antitumor effect of transferred T cells. *Proc Nat Acad Sci USA* 99, 16168–16173.

6. Butler, M.O., Friedlander, P., Milstein, M.I., Mooney, M.M., Metzler, G., Murray, A.P., Tanaka, M., Berezovskaya, A., Imataki, O., Drury, L., Brennan, L., Flavin, M., Neuberg, D., Stevenson, K., Lawrence, D., Hodi, F.S., Velazquez, E.F., Jaklitsch, M.T., Russell, S.E., Mihm, M., Nadler, L.M., and Hirano, N. Establishment of antitumor memory in humans using in vitro-educated CD8+ T cells. *Sci Transl Med* 3, 80ra34.

7. Middleton, M.R., Grob, J.J., Aaronson, N., Fierlbeck, G., Tilgen, W., Seiter, S., Gore, M., Aamdal, S., Cebon, J., Coates, A., Dreno, B., Henz, M., Schadendorf, D., Kapp, A., Weiss, J., Fraass, U., Statkevich, P., Muller, M., and Thatcher, N. (2000). Randomized phase III study of temozolomide versus dacarbazine in the treatment of patients with advanced metastatic malignant melanoma. *J Clin Oncol* 18, 158–166.

8. Rosenberg, S.A. and Dudley, M.E. (2009). Adoptive cell therapy for the treatment of patients with metastatic melanoma. *Curr Opin Immunol* 21, 233–240.

9. Boon, T., Coulie, P.G., Van den Eynde, B.J., and van der Bruggen, P. (2006). Human T cell responses against melanoma. *Annu Rev Immunol* 24, 175–208.

10. Brahmer, J.R., Drake, C.G., Wollner, I., Powderly, J.D., Picus, J., Sharfman, W.H., Stankevich, E., Pons, A., Salay, T.M., McMiller, T.L., Gilson, M.M., Wang, C., Selby, M., Taube, J.M., Anders, R., Chen, L., Korman, A.J., Pardoll, D.M., Lowy, I., and Topalian, S.L. Phase I study of single-agent anti-programmed death-1 (MDX-1106) in refractory solid tumors: Safety, clinical activity, pharmacodynamics, and immunologic correlates. *J Clin Oncol* 28, 3167–3175.

11. Hodi, F.S., O'Day, S.J., McDermott, D.F., Weber, R.W., Sosman, J.A., Haanen, J.B., Gonzalez, R., Robert, C., Schadendorf, D., Hassel, J.C., Akerley, W., van den Eertwegh, A.J., Lutzky, J., Lorigan, P., Vaubel, J.M., Linette, G.P., Hogg, D., Ottensmeier, C.H., Lebbe, C., Peschel, C., Quirt, I., Clark, J.I., Wolchok, J.D., Weber, J.S., Tian, J., Yellin, M.J., Nichol, G.M., Hoos, A., and Urba, W.J. Improved survival with ipilimumab in patients with metastatic melanoma. *N Engl J Med* 363, 711–723.

12. Coccoris, M., Straetemans, T., Govers, C., Lamers, C., Sleijfer, S., and Debetsm, R. T cell receptor (TCR) gene therapy to treat melanoma: Lessons from clinical and preclinical studies. *Expert Opin Biol Ther* 10, 547–562.

13. Linnemann, C., Schumacher, T.N., and Bendle, G.M. T-cell receptor gene therapy: Critical parameters for clinical success. *J Invest Dermatol* 131, 1806–1816.

14. Dembic, Z., Haas, W., Weiss, S., McCubrey, J., Kiefer, H., von Boehmer, H., and Steinmetz, M. (1986). Transfer of specificity by murine alpha and beta T-cell receptor genes. *Nature* 320, 232–238.

15. Kessels, H.W., Wolkers, M.C., van den Boom, M.D., van der Valk, M.A., and Schumacher, T.N. (2001). Immunotherapy through TCR gene transfer. *Nat Immunol* 2, 957–961.

16. Calogero, A., Hospers, G.A., Kruse, K.M., Schrier, P.I., Mulder, N.H., Hooijberg, E., and de Leij, L.F. (2000). Retargeting of a T cell line by anti MAGE-3/HLA-A2 alpha beta TCR gene transfer. *Anticancer Res* 20, 1793–1799.

17. Schaft, N., Willemsen, R.A., de Vries, J., Lankiewicz, B., Essers, B.W., Gratama, J.W., Figdor, C.G., Bolhuis, R.L., Debets, R., and Adema, G.J. (2003). Peptide fine specificity of anti-glycoprotein 100 CTL is preserved following transfer of engineered TCR alpha beta genes into primary human T lymphocytes. *J Immunol* 170, 2186–2194.

18. Morgan, R.A., Dudley, M.E., Wunderlich, J.R., Hughes, M.S., Yang, J.C., Sherry, R.M., Royal, R.E., Topalian, S.L., Kammula, U.S., Restifo, N.P., Zheng, Z., Nahvi, A., de Vries, C.R., Rogers-Freezer, L.J., Mavroukakis, S.A., and Rosenberg, S.A. (2006). Cancer regression in patients after transfer of genetically engineered lymphocytes. *Science* 314, 126–129.

19. Johnson, L.A., Morgan, R.A., Dudley, M.E., Cassard, L., Yang, J.C., Hughes, M.S., Kammula, U.S., Royal, R.E., Sherry, R.M., Wunderlich, J.R., Lee, C.C., Restifo, N.P., Schwarz, S.L., Cogdill, A.P., Bishop, R.J., Kim, H., Brewer, C.C., Rudy, S.F., VanWaes, C., Davis, J.L., Mathur, A., Ripley, R.T., Nathan, D.A., Laurencot, C.M., and Rosenberg, S.A. (2009). Gene therapy with human and mouse T-cell receptors mediates cancer regression and targets normal tissues expressing cognate antigen. *Blood* 114, 535–546.

20. Parkhurst, M.R., Yang, J.C., Langan, R.C., Dudley, M.E., Nathan, D.A., Feldman, S.A., Davis, J.L., Morgan, R.A., Merino, M.J., Sherry, R.M., Hughes, M.S., Kammula, U.S., Phan, G.Q., Lim, R.M., Wank, S.A., Restifo, N.P., Robbins, P.F., Laurencot, C.M., and Rosenberg, S.A. T cells targeting carcinoembryonic antigen can mediate regression of metastatic colorectal cancer but induce severe transient colitis. *Mol Ther* 19, 620–626.

21. Robbins, P.F., Morgan, R.A., Feldman, S.A., Yang, J.C., Sherry, R.M., Dudley, M.E., Wunderlich, J.R., Nahvi, A.V., Helman, L.J., Mackall, C.L., Kammula, U.S., Hughes, M.S., Restifo, N.P., Raffeld, M., Lee, C.C., Levy, C.L., Li, Y.F., El-Gamil, M., Schwarz, S.L., Laurencot, C., and Rosenberg, S.A. Tumor regression in patients with metastatic synovial cell sarcoma and melanoma using genetically engineered lymphocytes reactive with NY-ESO-1. *J Clin Oncol* 29, 917–924.

22. Overwijk, W.W., Theoret, M.R., Finkelstein, S.E., Surman, D.R., de Jong, L.A., Vyth-Dreese, F.A., Dellemijn, T.A., Antony, P.A., Spiess, P.J., Palmer,

D.C., Heimann, D.M., Klebanoff, C.A., Yu, Z., Hwang, L.N., Feigenbaum, L., Kruisbeek, A.M., Rosenberg, S.A., and Restifo, N.P. (2003). Tumor regression and autoimmunity after reversal of a functionally tolerant state of self-reactive CD8+ T cells. *J Exp Med* 198, 569–580.

23. Bos, R., van Duikeren, S., Morreau, H., Franken, K., Schumacher, T.N., Haanen, J.B., van der Burg, S.H., Melief, C.J., and Offringa, R. (2008). Balancing between antitumor efficacy and autoimmune pathology in T-cell-mediated targeting of carcinoembryonic antigen. *Cancer Res* 68, 8446–8455.

24. Ugel, S., Scarselli, E., Iezzi, M., Mennuni, C., Pannellini, T., Calvaruso, F., Cipriani, B., De Palma, R., Ricci-Vitiani, L., Peranzoni, E., Musiani, P., Zanovello, P., and Bronte, V. (2009). Autoimmune B cell lymphopenia following successful adoptive therapy with telomerase-specific T lymphocytes. *Blood*.

25. Palmer, D.C., Chan, C.C., Gattinoni, L., Wrzesinski, C., Paulos, C.M., Hinrichs, C.S., Powell, D.J., Klebanoff, Jr., C.A., Finkelstein, S.E., Fariss, R.N., Yu, Z., Nussenblatt, R.B., Rosenberg, S.A., and Restifo, N.P. (2008). Effective tumor treatment targeting a melanoma/melanocyte-associated antigen triggers severe ocular autoimmunity. *Proc Nat Acad Sci USA* 105, 8061–8066.

26. Dudley, M.E., Wunderlich, J.R., Robbins, P.F., Yang, J.C., Hwu, P., Schwartzentruber, D.J., Topalian, S.L., Sherry, R., Restifo, N.P., Hubicki, A.M., Robinson, M.R., Raffeld, M., Duray, P., Seipp, C.A., Rogers-Freezer, L., Morton, K.E., Mavroukakis, S.A., White, D.E., and Rosenberg, S.A. (2002). Cancer regression and autoimmunity in patients after clonal repopulation with antitumor lymphocytes. *Science* 298, 850–854.

27. Lamers, C.H., Sleijfer, S., Vulto, A.G., Kruit, W.H., Kliffen, M., Debets, R., Gratama, J.W., Stoter, G., and Oosterwijk, E. (2006). Treatment of metastatic renal cell carcinoma with autologous T-lymphocytes genetically retargeted against carbonic anhydrase IX: first clinical experience. *J Clin Oncol* 24, e20–e22.

28. Lucas, S. and Coulie, P.G. (2008). About human tumor antigens to be used in immunotherapy. *Semin Immunol* 20, 301–307.

29. Parmiani, G., De Filippo, A., Novellino, L., and Castelli, C. (2007). Unique human tumor antigens: Immunobiology and use in clinical trials. *J Immunol* 178, 1975–1979.

30. Chomez, P., De Backer, O., Bertrand, M., De Plaen, E., Boon, T., and Lucas, S. (2001). An overview of the MAGE gene family with the identification of all human members of the family. *Cancer Res* 61, 5544–5551.

31. Gotter, J., Brors, B., Hergenhahn, M., and Kyewski, B. (2004). Medullary epithelial cells of the human thymus express a highly diverse selection of tissue-specific genes colocalized in chromosomal clusters. *J Exp Med* 199, 155–166.

32. Caballero, O.L. and Chen, Y.T. (2009). Cancer/testis (CT) antigens: Potential targets for immunotherapy. *Cancer Sci* 100, 2014–2021.

33. van der Bruggen, P., Traversari, C., Chomez, P., Lurquin, C., De Plaen, E., Van den Eynde, B., Knuth, A., and Boon, T. (1991). A gene encoding an antigen recognized by cytolytic T lymphocytes on a human melanoma. *Science* 254, 1643–1647.

34. Yang, B., O'Herrin, S.M., Wu, J., Reagan-Shaw, S., Ma, Y., Bhat, K.M., Gravekamp, C., Setaluri, V., Peters, N., Hoffmann, F.M., Peng, H., Ivanov, A.V., Simpson, A.J., and Longley, B.J. (2007). MAGE-A, mMage-b, and MAGE-C proteins form complexes with KAP1 and suppress p53-dependent apoptosis in MAGE-positive cell lines. *Cancer Res* 67, 9954–9962.

35. Liu, W., Cheng, S., Asa, S.L., and Ezzat, S. (2008). The melanoma-associated antigen A3 mediates fibronectin-controlled cancer progression and metastasis. *Cancer Res* 68, 8104–8112.

36. Sigalotti, L., Covre, A., Zabierowski, S., Himes, B., Colizzi, F., Natali, P.G., Herlyn, M., and Maio, M. (2008). Cancer testis antigens in human melanoma stem cells: Expression, distribution, and methylation status. *J Cell Physiol* 215, 287–291.

37. Willemsen, R.A., Weijtens, M.E., Ronteltap, C., Eshhar, Z., Gratama, J.W., Chames, P., and Bolhuis, R.L. (2000). Grafting primary human T lymphocytes with cancer-specific chimeric single chain and two chain TCR. *Gene Ther* 7, 1369–1377.

38. Zhao, Y., Zheng, Z., Robbins, P.F., Khong, H.T., Rosenberg, S.A., and Morgan, R.A. (2005). Primary human lymphocytes transduced with NY-ESO-1 antigen-specific TCR genes recognize and kill diverse human tumor cell lines. *J Immunol* 174, 4415–4423.

39. Zhao, Y., Zheng, Z., Khong, H.T., Rosenberg, S.A., and Morgan, R.A. (2006). Transduction of an HLA-DP4-restricted NY-ESO-1-specific TCR into primary human CD4+ lymphocytes. *J Immunother* 29, 398–406.

40. Spiotto, M.T., Rowley, D.A., and Schreiber, H. (2004). Bystander elimination of antigen loss variants in established tumors. *Nat Med* 10, 294–298.

41. Engels, B., Rowley, D.A., and Schreiber, H. Targeting stroma to treat cancers. *Semin Cancer Biol* 22, 41–49.

42. Ning, N., Pan, Q. Zheng, F., Teitz-Tennenbaum, S., Egenti, M., Yet, J., Li, M., Ginestier, C., Wicha, M.S., Moyer, J.S., Prince, M.E., Xu, Y., Zhang, X.L., Huang, S., Chang, A.E., and Li, Q. Cancer stem cell vaccination confers significant antitumor immunity. *Cancer Res* 72, 1853–1864.

43. Bendle, G.M., Linnemann, C., Hooijkaas, A.I., Bies, L., de Witte, M.A., Jorritsma, A., Kaiser, A.D., Pouw, N., Debets, R., Kieback, E., Uckert, W., Song, J.Y., Haanen, J.B., and Schumacher, T.N. Lethal graft-versus-host disease in mouse models of T cell receptor gene therapy. *Nat Med* 16, 565–570.

44. van Loenen, M.M., de Boer, R., Amir, A.L., Hagedoorn, R.S., Volbeda, G.L., Willemze, R., van Rood, J.J., Falkenburg, J.H., and Heemskerk, M.H. Mixed

T cell receptor dimers harbor potentially harmful neoreactivity. *Proc Natl Acad Sci USA* 107, 10972–10977.

45. Govers, C., Sebestyen, Z., Coccoris, M., Willemsen, R.A., and Debets, R. T cell receptor gene therapy: Strategies for optimizing transgenic TCR pairing. *Trends Mol Med* 16, 77–87.

46. Cohen, C.J., Zhao, Y., Zheng, Z., Rosenberg, S.A., and Morgan, R.A. (2006). Enhanced antitumor activity of murine-human hybrid T-cell receptor (TCR) in human lymphocytes is associated with improved pairing and TCR/CD3 stability. *Cancer Res* 66, 8878–8886.

47. Voss, R.H., Kuball, J., Engel, R., Guillaume, P., Romero, P., Huber, C., and Theobald, M. (2006). Redirection of T cells by delivering a transgenic mouse-derived MDM2 tumor antigen-specific TCR and its humanized derivative is governed by the CD8 coreceptor and affects natural human TCR expression. *Immunol. Res* 34, 67–87.

48. Thomas, S., Xue, S.A., Cesco-Gaspere, M., San Jose, E., Hart, D.P., Wong, V., Debets, R., Alarcon, B., Morris, E., and Stauss, H.J. (2007). Targeting the Wilms tumor antigen 1 by TCR gene transfer: TCR variants improve tetramer binding but not the function of gene modified human T cells. *J Immunol* 179, 5803–5810.

49. Sommermeyer, D. and Uckert, W. Minimal amino acid exchange in human TCR constant regions fosters improved function of TCR gene-modified T cells. *J Immunol* 184, 6223–6231.

50. Bialer, G., Horovitz-Fried, M., Ya'acobi, S., Morgan, R.A., and Cohen, C.J. Selected murine residues endow human TCR with enhanced tumor recognition. *J Immunol* 184, 6232–6241.

51. Boulter, J.M., Glick, M., Todorov, P.T., Baston, E., Sami, M., Rizkallah, P., and Jakobsen, B.K. (2003). Stable, soluble T-cell receptor molecules for crystallization and therapeutics. *Protein Eng* 16, 707–711.

52. Kuball, J., Dossett, M.L., Wolfl, M., Ho, W.Y., Voss, R.H., Fowler, C., and Greenberg, P.D. (2007). Facilitating matched pairing and expression of TCR chains introduced into human T cells. *Blood* 109, 2331–2338.

53. Cohen, C.J., Li, Y.F., El-Gamil, M., Robbins, P.F., Rosenberg, S.A., and Morgan, R.A. (2007). Enhanced antitumor activity of T cells engineered to express T-cell receptors with a second disulfide bond. *Cancer Res* 67, 3898–3903.

54. Voss, R.H., Willemsen, R.A., Kuball, J., Grabowski, M., Engel, R., Intan, R.S., Guillaume, P., Romero, P., Huber, C., and Theobald, M. (2008). Molecular design of the Calphabeta interface favors specific pairing of introduced TCRalphabeta in human T cells. *J Immunol* 180, 391–401.

55. Sebestyen, Z., Schooten, E., Sals, T., Zaldivar, I., San Jose, E., Alarcon, B., Bobisse, S., Rosato, A., Szollosi, J., Gratama, J.W., Willemsen, R.A., and Debets, R. (2008). Human TCR that incorporate CD3zeta induce highly

preferred pairing between TCRalpha and beta chains following gene transfer. *J Immunol* 180, 7736–7746.

56. Schaft, N., Lankiewicz, B., Gratama, J.W., Bolhuis, R.L., and Debets, R. (2003). Flexible and sensitive method to functionally validate tumor-specific receptors via activation of NFAT. *J Immunol Methods* 280, 13–24.

57. Chung, S., Wucherpfennig, K.W., Friedman, S.M., Hafler, D.A., and Strominger, J.L. (1994). Functional three-domain single-chain T-cell receptors. *Proc Nat Acad Sci USA* 91, 12654–12658.

58. Aggen, D.H., Chervin, A.S., Schmitt, T.M., Engels, B., Stone, J.D., Richman, S.A., Piepenbrink, K.H., Baker, B.M., Greenberg, P.D., Schreiber, H., and Kranz, D.M. Single-chain ValphaVbeta T-cell receptors function without mispairing with endogenous TCR chains. *Gene Ther* 19, 365–374.

59. Voss, R.H., Thomas, S., Pfirschke, C., Hauptrock, B., Klobuch, S., Kuball, J., Grabowski, M., Engel, R., Guillaume, P., Romero, P., Huber, C., Beckhove, P., and Theobald, M. Coexpression of the T-cell receptor constant alpha domain triggers tumor reactivity of single-chain TCR-transduced human T cells. *Blood* 115, 5154–5163.

60. Okamoto, S., Mineno, J., Ikeda, H., Fujiwara, H., Yasukawa, M., Shiku, H., and Kato, I. (2009). Improved expression and reactivity of transduced tumor-specific TCRs in human lymphocytes by specific silencing of endogenous TCR. *Cancer Res* 69, 9003–9011.

61. Provasi, E., Genovese, P., Lombardo, A., Magnani, Z., Liu, P.Q., Reik, A., Chu, V., Paschon, D.E., Zhang, L., Kuball, J., Camisa, B., Bondanza, A., Casorati, G., Ponzoni, M., Ciceri, F., Bordignon, C., Greenberg, P.D., Holmes, M.C., Gregory, P.D., Naldini, L., and Bonini, C. Editing T cell specificity towards leukemia by zinc finger nucleases and lentiviral gene transfer. *Nat Med*.

62. Heemskerk, M.H., Hoogeboom, M., Hagedoorn, R., Kester, M.G., Willemze, R. and Falkenburg, J.H. (2004). Reprogramming of virus-specific T cells into leukemia-reactive T cells using T cell receptor gene transfer. *J Exp Med* 199, 885–894.

63. van der Veken, L.T., Coccoris, M., Swart, E., Falkenburg, J.H., Schumacher, T.N., and Heemskerk, M.H. (2009). Alpha beta T cell receptor transfer to gamma delta T cells generates functional effector cells without mixed TCR dimers in vivo. *J Immunol* 182, 164–170.

64. Bonini, C., Ferrari, G., Verzeletti, S., Servida, P., Zappone, E., Ruggieri, L., Ponzoni, M., Rossini, S., Mavilio, F., Traversari, C., and Bordignon, C. (1997). HSV-TK gene transfer into donor lymphocytes for control of allogeneic graft-versus-leukemia. *Science* 276, 1719–1724.

65. de Witte, M.A., Jorritsma, A., Swart, E., Straathof, K.C., de Punder, K., Haanen, J.B., Rooney, C.M., and Schumacher, T.N. (2008). An inducible

caspase 9 safety switch can halt cell therapy-induced autoimmune disease. *J Immunol* 180, 6365–6373.

66. Straathof, K.C., Pule, M.A., Yotnda, P., Dotti, G., Vanin, E.F., Brenner, M.K., Heslop, H.E., Spencer, D.M., and Rooney, C.M. (2005). An inducible caspase 9 safety switch for T-cell therapy. *Blood* 105, 4247–4254.

67. Serafini, M., Manganini, M., Borleri, G., Bonamino, M., Imberti, L., Biondi, A., Golay, J., Rambaldi, A., and Introna, M. (2004). Characterization of CD20-transduced T lymphocytes as an alternative suicide gene therapy approach for the treatment of graft-versus-host disease. *Hum Gene Ther* 15, 63–76.

68. Kieback, E., Charo, J., Sommermeyer, D., Blankenstein, T., and Uckert, W. (2008). A safeguard eliminates T cell receptor gene-modified autoreactive T cells after adoptive transfer. *Proc Natl Acad Sci USA* 105, 623–628.

69. Pouw, N.M., Westerlaken, E.J., Willemsen, R.A., and Debets, R. (2007). Gene transfer of human TCR in primary murine T cells is improved by pseudo-typing with amphotropic and ecotropic envelopes. *J Gene Med* 9, 561–570.

70. Lamers, C.H., Willemsen, R.A., van Elzakker, P., van Krimpen, B.A., Gratama, J.W., and Debets, R. (2006). Phoenix-ampho outperforms PG13 as retroviral packaging cells to transduce human T cells with tumor-specific receptors: Implications for clinical immunogene therapy of cancer. *Cancer Gene Ther* 13, 503–509.

71. Jones, S., Peng, P.D., Yang, S., Hsu, C., Cohen, C.J., Zhao, Y., Abad, J., Zheng, Z., Rosenberg, S.A., and Morgan, R.A. (2009). Lentiviral vector design for optimal T cell receptor gene expression in the transduction of peripheral blood lymphocytes and tumor-infiltrating lymphocytes. *Hum Gene Ther* 20, 630–640.

72. Schambach, A., Wodrich, H., Hildinger, M., Bohne, J., Krausslich, H.G., and Baum, C. (2000). Context dependence of different modules for posttranscriptional enhancement of gene expression from retroviral vectors. *Mol Ther* 2, 435–445.

73. Engels, B., Cam, H., Schuler, T., Indraccolo, S., Gladow, M., Baum, C., Blankenstein, T., and Uckert, W. (2003). Retroviral vectors for high-level transgene expression in T lymphocytes. *Hum Gene Ther* 14, 1155–1168.

74. Mizuguchi, H., Xu, Z., Ishii-Watabe, A., Uchida, E., and Hayakawa, T. (2000). IRES-dependent second gene expression is significantly lower than cap-dependent first gene expression in a bicistronic vector. *Mol Ther* 1, 376–382.

75. de Felipe, P., Martin, V.M, Cortes, M.L., Ryan, M., and Izquierdo, M. (1999). Use of the 2A sequence from foot-and-mouth disease virus in the generation of retroviral vectors for gene therapy. *Gene Ther* 6, 198–208.

76. Klump, H., Schiedlmeier, B., Vogt, B., Ryan, M., Ostertag, W., and Baum, C. (2001). Retroviral vector-mediated expression of HoxB4 in hematopoietic cells using a novel coexpression strategy. *Gene Ther* 8, 811–817.

77. Leisegang, M., Engels, B., Meyerhuber, P., Kieback, E., Sommermeyer, D., Xue, S.A., Reuss, S., Stauss, H., and Uckert, W. (2008). Enhanced functionality of T cell receptor-redirected T cells is defined by the transgene cassette. *J Mol Med* (Berl) 86, 573–583.

78. Hacein-Bey-Abina, S., Von Kalle, C., Schmidt, M., McCormack, M.P., Wulffraat, N., Leboulch, P., Lim, A., Osborne, C.S., Pawliuk, R., Morillon, E., Sorensen, R., Forster, A., Fraser, P., Cohen, J.I., de Saint Basile, G., Alexander, I., Wintergerst, U., Frebourg, T., Aurias, A., Stoppa-Lyonnet, D., Romana, S., Radford-Weiss, I., Gross, F., Valensi, F., Delabesse, E., Macintyre, E., Sigaux, F., Soulier, J., Leiva, L.E., Wissler, M., Prinz, C., Rabbitts, T.H., Le Deist, F., Fischer, A., and Cavazzana-Calvo, M. (2003). LMO2-associated clonal T cell proliferation in two patients after gene therapy for SCID-X1. *Science* 302, 415–419.

79. Newrzela, S., Cornils, K., Li, Z., Baum, C., Brugman, M.H., Hartmann, M., Meyer, J., Hartmann, S., Hansmann, M.L., Fehse, B., and von Laer, D. (2008). Resistance of mature T cells to oncogene transformation. *Blood* 112, 2278–2286.

80. Scholler, J., Brady, T.L., Binder-Scholl, G., Hwang, W.T., Plesa, G., Hege, K.M., Vogel, A.N., Kalos, M., Riley, J.L., Deeks, S.G., Mitsuyasu, R.T., Bernstein, W.B., Aronson, N.E., Levine, B.L., Bushman, F.D., and June, C.H. Decade-long safety and function of retroviral-modified chimeric antigen receptor T cells. *Sci Transl Med* 4, 132ra153.

81. Ross, J. (1995). mRNA stability in mammalian cells. *Microbiol Rev* 59, 423–450.

82. Jorritsma, A., Gomez-Eerland, R., Dokter, M., van de Kasteele, W., Zoet, Y.M., Doxiadis, I.I., Rufer, N., Romero, P., Morgan, R.A., Schumacher, T.N., and Haanen, J.B. (2007). Selecting highly affine and well-expressed TCRs for gene therapy of melanoma. *Blood* 110, 3564–3572.

83. Scholten, K.B., Kramer, D., Kueter, E.W., Graf, M., Schoedl, T., Meijer, C.J., Schreurs, M.W. and Hooijberg, E. (2006). Codon modification of T cell receptors allows enhanced functional expression in transgenic human T cells. *Clin Immunol* 119, 135–145.

84. Haga-Friedman, A., Horovitz-Fried, M., and Cohen, C.J. Incorporation of transmembrane hydrophobic mutations in the TCR enhance its surface expression and T cell functional avidity. *J Immunol* 188, 5538–5546.

85. Kuball, J., Hauptrock, B., Malina, V., Antunes, E., Voss, R.H., Wolfl, M., Strong, R., Theobald, M., and Greenberg, P.D. (2009). Increasing functional avidity of TCR-redirected T cells by removing defined N-glycosylation sites in the TCR constant domain. *J Exp Med* 206, 463–475.

86. Varela-Rohena, A., Molloy, P.E., Dunn, S.M., Li, Y., Suhoski, M.M., Carroll, R.G., Milicic, A., Mahon, T., Sutton, D.H., Laugel, B., Moysey, R., Cameron, B.J., Vuidepot, A., Purbhoo, M.A., Cole, D.K., Phillips, R.E., June, C.H.,

Jakobsen, B.K., Sewell, A.K., and Riley, J.L. (2008). Control of HIV-1 immune escape by CD8 T cells expressing enhanced T-cell receptor. *Nat Med* 14, 1390–1395.

87. Robbins, P.F., Li, Y.F., El-Gamil, M., Zhao, Y., Wargo, J.A., Zheng, Z., Xu, H., Morgan, R.A., Feldman, S.A., Johnson, L.A., Bennett, A.D., Dunn, S.M., Mahon, T.M., Jakobsen, B.K., and Rosenberg, S.A. (2008). Single and dual amino acid substitutions in TCR CDRs can enhance antigen-specific T cell functions. *J Immunol* 180, 6116–6131.

88. Sadovnikova, E., Jopling, L.A., Soo, K.S., and Stauss, H.J. (1998). Generation of human tumor-reactive cytotoxic T cells against peptides presented by non-self HLA class I molecules. *Eur J Immunol* 28, 193–200.

89. Davis, J.L., Theoret, M.R., Zheng, Z., Lamers, C.H., Rosenberg, S.A., and Morgan, R.A. Development of human anti-murine T-cell receptor antibodies in both responding and nonresponding patients enrolled in TCR gene therapy trials. *Clin Cancer Res* 16, 5852–5861.

90. Li, L.P., Lampert, J.C., Chen, X., Leitao, C., Popovic, J., Muller, W., and Blankenstein, T. Transgenic mice with a diverse human T cell antigen receptor repertoire. *Nat Med* 16, 1029–1034.

91. Ahmadi, M., King, J.W., Xue, S.A., Voisine, C., Holler, A., Wright, G.P., Waxman, J., Morris, E., and Stauss, H.J. CD3 limits the efficacy of TCR gene therapy in vivo. *Blood* 118, 3528–3537.

92. Geiger, T.L., Nguyen, P., Leitenberg, D., and Flavell, R.A. (2001). Integrated src kinase and costimulatory activity enhances signal transduction through single-chain chimeric receptors in T lymphocytes. *Blood* 98, 2364–2371.

93. Zhang, T., He, X., Tsang, T.C., and Harris, D.T. (2004). Transgenic TCR expression: Comparison of single chain with full-length receptor constructs for T-cell function. *Cancer Gene Ther* 11, 487–496.

94. Schaft, N., Lankiewicz, B., Drexhage, J., Berrevoets, C., Moss, D.J., Levitsky, V., Bonneville, M., Lee, S.P., McMichael, A.J., Gratama, J.W., Bolhuis, R.L., Willemsen, R., and Debets, R. (2006). T cell re-targeting to EBV antigens following TCR gene transfer: CD28-containing receptors mediate enhanced antigen-specific IFNgamma production. *Int Immunol* 18, 591–601.

95. Willemsen, R.A., Ronteltap, C., Chames, P., Debets, R., and Bolhuis, R.L. (2005). T cell retargeting with MHC class I-restricted antibodies: The CD28 costimulatory domain enhances antigen-specific cytotoxicity and cytokine production. *J Immunol* 174, 7853–7858.

96. Hombach, A. and Abken, H. (2007). Costimulation tunes tumor-specific activation of redirected T cells in adoptive immunotherapy. *Cancer Immunol Immunother* 56, 731–737.

97. Carpenito, C., Milone, M.C., Hassan, R., Simonet, J.C., Lakhal, M., Suhoski, M.M., Varela-Rohena, A., Haines, K.M., Heitjan, D.F., Albelda, S.M., Carroll,

R.G., Riley, J.L., Pastan, I., and June, C.H. (2009). Control of large, established tumor xenografts with genetically retargeted human T cells containing CD28 and CD137 domains. *Proc Natl Acad Sci USA* 106, 3360–3365.

98. Schietinger, A., Philip, M., Liu, R.B., Schreiber, K., and Schreiber, H. Bystander killing of cancer requires the cooperation of CD4(+) and CD8(+) T cells during the effector phase. *J Exp Med* 207, 2469–2477.

99. Dudley, M.E., Wunderlich, J.R., Yang, J.C., Sherry, R.M., Topalian, S.L., Restifo, N.P., Royal, R.E., Kammula, U., White, D.E., Mavroukakis, S.A., Rogers, L.J., Gracia, G.J., Jones, S.A., Mangiameli, D.P., Pelletier, M.M., Gea-Banacloche, J., Robinson, M.R., Berman, D.M., Filie, A.C., Abati, A., and Rosenberg, S.A. (2005). Adoptive cell transfer therapy following non-myeloablative but lymphodepleting chemotherapy for the treatment of patients with refractory metastatic melanoma. *J Clin Oncol* 23, 2346–2357.

100. Huang, J., Khong, H.T., Dudley, M.E., El-Gamil, M., Li, Y.F., Rosenberg, S.A., and Robbins, P.F. (2005). Survival, persistence, and progressive differentiation of adoptively transferred tumor-reactive T cells associated with tumor regression. *J Immunother* 28, 258–267.

101. Kaneko, S., Mastaglio, S., Bondanza, A., Ponzoni, M., Sanvito, F., Aldrighetti, L., Radrizzani, M., La Seta-Catamancio, S., Provasi, E., Mondino, A., Nagasawa, T., Fleischhauer, K., Russo, V., Traversari, C., Ciceri, F., Bordignon, C., and Bonini, C. (2009). IL-7 and IL-15 allow the generation of suicide gene-modified alloreactive self-renewing central memory human T lymphocytes. *Blood* 113, 1006–1015.

102. Pouw, N.T.-W.E., Kraan, J., Jonker, M., Wittink, F., ten Hagen, T., Verweij, J., and Debets, R. Antigen-specific production of TCR-redirected primary T cells strongly enhanced by pretreatment with a combination of IL-15 and IL-21. *Cancer Immunol Immunother* (in press).

103. Berger, C., Jensen, M.C., Lansdorp, P.M., Gough, M., Elliott, C., and Riddell, S.R. (2008). Adoptive transfer of effector CD8+ T cells derived from central memory cells establishes persistent T cell memory in primates. *J Clin Invest* 118, 294–305.

104. Hinrichs, C.S., Borman, Z.A., Cassard, L., Gattinoni, L., Spolski, R., Yu, Z., Sanchez-Perez, L., Muranski, P., Kern, S.J., Logun, C., Palmer, D.C., Ji, Y., Reger, R.N., Leonard, W.J., Danner, R.L., Rosenberg, S.A., and Restifo, N.P. (2009). Adoptively transferred effector cells derived from naive rather than central memory CD8+ T cells mediate superior antitumor immunity. *Proc Natl Acad Sci USA* 106, 17469–17474.

105. Gattinoni, L., Lugli, E., Ji, Y., Pos, Z., Paulos, C.M., Quigley, M.F., Almeida, J.R., Gostick, E., Yu, Z., Carpenito, C., Wang, E., Douek, D.C., Price, D.A., June, C.H., Marincola, F.M., Roederer, M., and Restifo, N.P. A human memory T cell subset with stem cell-like properties. *Nat Med* 17, 1290–1297.

106. Hunziker, L., Klenerman, P., Zinkernagel, R.M., and Ehl, S. (2002). Exhaustion of cytotoxic T cells during adoptive immunotherapy of virus carrier mice can be prevented by B cells or CD4+ T cells. *Eur J Immunol* 32, 374–382.

107. Marzo, A.L., Kinnear, B.F., Lake, R.A., Frelinger, J.J., Collins, E.J., Robinson, B.W., and Scott, B. (2000). Tumor-specific CD4+ T cells have a major "post-licensing" role in CTL mediated anti-tumor immunity. *J Immunol* 165, 6047–6055.

108. Antony, P.A., Piccirillo, C.A., Akpinarli, A., Finkelstein, S.E., Speiss, P.J., Surman, D.R., Palmer, D.C., Chan, C.C., Klebanoff, C.A., Overwijk, W.W., Rosenberg, S.A., and Restifo, N.P. (2005). CD8+ T cell immunity against a tumor/self-antigen is augmented by CD4+ T helper cells and hindered by naturally occurring T regulatory cells. *J Immunol* 174, 2591–2601.

109. Gyobu, H., Tsuji, T., Suzuki, Y., Ohkuri, T., Chamoto, K., Kuroki, M., Miyoshi, H., Kawarada, Y., Katoh, H., Takeshima, T., and Nishimura, T. (2004). Generation and targeting of human tumor-specific Tc1 and Th1 cells transduced with a lentivirus containing a chimeric immunoglobulin T-cell receptor. *Cancer Res* 64, 1490–1495.

110. Nishimura, T., Iwakabe, K., Sekimoto, M., Ohmi, Y., Yahata, T., Nakui, M., Sato, T., Habu, S., Tashiro, H., Sato, M., and Ohta, A. (1999). Distinct role of antigen-specific T helper type 1 (Th1) and Th2 cells in tumor eradication in vivo. *J Exp Med* 190, 617–627.

111. Muranski, P., Boni, A., Antony, P.A., Cassard, L., Irvine, K.R., Kaiser, A., Paulos, C.M., Palmer, D.C., Touloukian, C.E., Ptak, K., Gattinoni, L., Wrzesinski, C., Hinrichs, C.S., Kerstann, K.W., Feigenbaum, L., Chan, C.C., and Restifo, N.P. (2008). Tumor-specific Th17-polarized cells eradicate large established melanoma. *Blood* 112, 362–373.

112. Kuball, J., Schmitz, F.W., Voss, R.H., Ferreira, E.A., Engel, R., Guillaume, P., Strand, S., Romero, P., Huber, C., Sherman, L.A., and Theobald, M. (2005). Cooperation of human tumor-reactive CD4+ and CD8+ T cells after redirection of their specificity by a high-affinity p53A2.1-specific TCR. *Immunity* 22, 117–129.

113. Willemsen, R., Ronteltap, C., Heuveling, M., Debets, R., and Bolhuis, R. (2005). Redirecting human CD4+ T lymphocytes to the MHC class I-restricted melanoma antigen MAGE-A1 by TCR alphabeta gene transfer requires CD8alpha. *Gene Ther* 12, 140–146.

114. Willemsen, R.A., Sebestyen, Z., Ronteltap, C., Berrevoets, C., Drexhage, J., and Debets, R. (2006). CD8 alpha coreceptor to improve TCR gene transfer to treat melanoma: Down-regulation of tumor-specific production of IL-4, IL-5, and IL-10. *J Immunol* 177, 991–998.

115. Morris, E.C., Tsallios, A., Bendle, G.M., Xue, S.A., and Stauss, H.J. (2005). A critical role of T cell antigen receptor-transduced MHC class I-restricted helper T cells in tumor protection. *Proc Natl Acad Sci USA* 102, 7934–7939.

116. Muranski, P., Borman, Z.A., Kerkar, S.P., Klebanoff, C.A., Ji, Y., Sanchez-Perez, L., Sukumar, M., Reger, R.N., Yu, Z., Kern, S.J., Roychoudhuri, R., Ferreyra, G.A., Shen, W., Durum, S.K., Feigenbaum, L., Palmer, D.C., Antony, P.A., Chan, C.C., Laurence, A., Danner, R.L., Gattinoni, L., and Restifo, N.P. Th17 cells are long lived and retain a stem cell-like molecular signature. *Immunity* 35, 972–985.

117. Rossig, C., Bollard, C.M., Nuchtern, J.G., Rooney, C.M., and Brenner, M.K. (2002). Epstein-Barr virus-specific human T lymphocytes expressing antitumor chimeric T-cell receptors: Potential for improved immunotherapy. *Blood* 99, 2009–2016.

118. Pule, M.A., Savoldo, B., Myers, G.D., Rossig, C., Russell, H.V., Dotti, G., Huls, M.H., Liu, E., Gee, A.P., Mei, Z., Yvon, E., Weiss, H.L., Liu, H., Rooney, C.M., Heslop, H.E., and Brenner, M.K. (2008). Virus-specific T cells engineered to coexpress tumor-specific receptors: Persistence and antitumor activity in individuals with neuroblastoma. *Nat Med* 14, 1264–1270.

119. Teague, R.M., Greenberg, P.D., Fowler, C., Huang, M.Z., Tan, X., Morimoto, J., Dossett, M.L., Huseby, E.S., and Ohlen, C. (2008). Peripheral CD8+ T cell tolerance to self-proteins is regulated proximally at the T cell receptor. *Immunity* 28, 662–674.

120. Hawkins, R.E., Gilham, D.E., Debets, R., Eshhar, Z., Taylor, N., Abken, H., Schumacher, T.N., and ATTACK Consortium. Development of adoptive cell therapy for cancer: A clinical perspective. *Hum Gene Ther* 21, 665–672.

121. Chen, J.L., Dunbar, P.R., Gileadi, U., Jager, E., Gnjatic, S., Nagata, Y., Stockert, E., Panicali, D.L., Chen, Y.T., Knuth, A., Old, L.J., and Cerundolo, V. (2000). Identification of NY-ESO-1 peptide analogues capable of improved stimulation of tumor-reactive CTL. *J Immunol* 165, 948–955.

122. Zhao, Y., Bennett, A.D., Zheng, Z., Wang, Q.J., Robbins, P.F., Yu, L.Y., Li, Y., Molloy, P.E., Dunn, S.M., Jakobsen, B.K., Rosenberg, S.A., and Morgan, R.A. (2007). High-affinity TCRs generated by phage display provide CD4+ T cells with the ability to recognize and kill tumor cell lines. *J Immunol* 179, 5845–5854.

123. Toebes, M., Coccoris, M., Bins, A., Rodenko, B., Gomez, R., Nieuwkoop, N.J., van de Kasteele, W., Rimmelzwaan, G.F., Haanen, J.B., Ovaa, H., and Schumacher, T.N. (2006). Design and use of conditional MHC class I ligands. *Nat Med* 12, 246–251.

124. Hadrup, S.R., Bakker, A.H., Shu, C.J., Andersen, R.S., van Veluw, J., Hombrink, P., Castermans, E., Thor Straten, P., Blank, C., Haanen, J.B., Heemskerk, M.H., and Schumacher, T.N. (2009). Parallel detection of antigen-specific T-cell responses by multidimensional encoding of MHC multimers. *Nat Methods* 6, 520–526.

125. Gilham, D.E., Debets, R., Pule, M., Hawkins, R.E., and Abken, H. CAR-T cells and solid tumors: Tuning T cells to challenge an inveterate foe. *Trends Mol Med*.

Chapter 4

T-Bodies: Antibody-Based Engineered T-Cell Receptors

John Bridgeman[1], Andreas A. Hombach[2], David Gilham[1],
Zelig Eshhar[3] and Hinrich Abken[2]

[1]Cellular Therapy Group, Cancer Research UK Department
of Medical Oncology, University of Manchester, UK
[2]Centre for Molecular Medicine Cologne and Clinic I Internal Medicine,
University of Cologne, Cologne, Germany
[3]Weizmann Institute for Sciences, Rehovot, Israel

4.1 Overview

Redirecting T cells by recombinant targeting receptor molecules has attracted increasing interest for use in the adoptive immunotherapy of malignant diseases. Difficulties in engrafting both recombinant chains of the T-cell receptor (TCR) in T cells, the potential of mis-pairing of recombinant TCR chains with the endogenous physiological TCR, and the frequent defects in antigen processing and presentation on tumor cells led to the development of a chimeric, MHC-independent recombinant TCR whose antigen-binding domain is derived from an antibody and signaling domain from the TCR. The archetypal chimeric antigen receptor (CAR; immunoreceptor), nick-named "T-body", consists of one polypeptide chain with an extracellular domain for specific antigen binding, typically a single chain fragment of variable region (scFv) antibody, which is fused to the transmembrane (TM) and cytoplasmic domain with signaling moieties, frequently the CD3ζ of the TCR or the γ-chain of the immunoglobulin FcεRI receptor. Due to the single polypeptide chain configuration and the modular composition, the CAR design substantially simplifies expression and function of the recombinant receptor in T cells. Upon antigen engagement, the intracellular signaling domain initiates T-cell activation resulting in T-cell amplification, cytokine secretion, and specific cytolysis of antigen-positive target cells. CARs overcome many limitations of TCRs being MHC unrestricted, thus recognizing antigens shared by many individuals independently of their MHC. Whereas the technical procedures to engineer redirected T cells are well established, optimization of the CAR structure has been a long-standing aspect of research. In most cases, optimization strategies have focused on the addition of co-stimulatory moieties to the cytoplasmic signaling domain; however, data suggest that modifications to extracellular and TM domains also have significant impact on CAR function. In this chapter, we summarize the structural prerequisites of CARs for efficient T-cell activation, the benefits of CARs with an integrated co-stimulatory domain, and the most recent development of CARs with combined co-stimulatory domains. The clinical potential as well as ongoing clinical trials using CAR-modified T cells in the therapy of malignant diseases are discussed.

4.2 A Long Way to the One-Chain Format: A Brief History of T-Bodies

The first attempts to confer on T-cells antibody-type specificity took advantage of the similarity in structure and genetic organization of the TCR and antibody and replaced the Vα and Vβ fragments of the TCR with either one of the V$_H$ and V$_L$ fragments of a given antibody.[1] Upon expression in T cells, such two-chain chimeric receptor genes form heterodimers, associated with the CD3 complex and proved functional in signaling for all the effector functions of T cells.[2,3] T cells expressing the antibody-based chimeric receptors, nick-named "T-bodies", were primarily used to study physico-chemical parameters of T-cell interactions under non-MHC-restricted conditions. Due to the inefficiency of the T-cell transfection technology at that time and the emerging potential of the use of antigen-specific T cells for cancer immunotherapy, a scFv antibody, that was proven to entail the antibody binding activity,[4,5] was directly linked to lymphocyte signaling molecules, such as the CD3 ζ-chain or the FcεRI γ-chain, as a single chain chimeric receptor.[6] This second generation of CAR together with the development of efficient and safe technologies of gene transfer to T cells using γ-retroviral vectors have become one of the major methods of choice to redirect the specificity of T cells. The modular structure of the scFv-CAR enabled the further additions of T-cell co-stimulatory domains[7–9] to the cytoplasmatic part of the CAR thus endowing T cells expressing such third generation of CAR with full T-cell activation potential, independent of MHC, and co-stimulatory ligand expression on its target cells.

4.3 From Structure to Function

Due to the modular composition, the individual domains, i.e., the extracellular binding domain, the spacer domain, the TM part, and the intracellular signaling domain, can be replaced by others to alter specificity and function of the CAR (Figure 4.1). Efficient gene transfer protocols have been developed during the last few years to engineer CAR-modified T cells in sufficient numbers for clinical use, retro- and lentiviral gene transfer has been most frequently used while RNA transfer was recently reported (see Chapter 2). Regardless of the gene delivery method, the CAR structure itself has a major impact on both the expression on T cells and the function of redirected T cells, a factor which is discussed below.

4.3.1 *The binding domain*

The modular composition of CARs makes it easy to change the specificity by substitution of the scFv antibody domain. Recent advantages in recombinant antibody technologies have led to increasing numbers of single chain antibodies (scFv) and a number of CARs with specificity for a variety of target antigens have been generated over the last few years, mostly against tumor-associated antigens and also against virus-infected cells (Table 4.1). Whereas some target antigens are polypeptides, others are highly glycosylated peptide antigens or

Bridgeman et al.

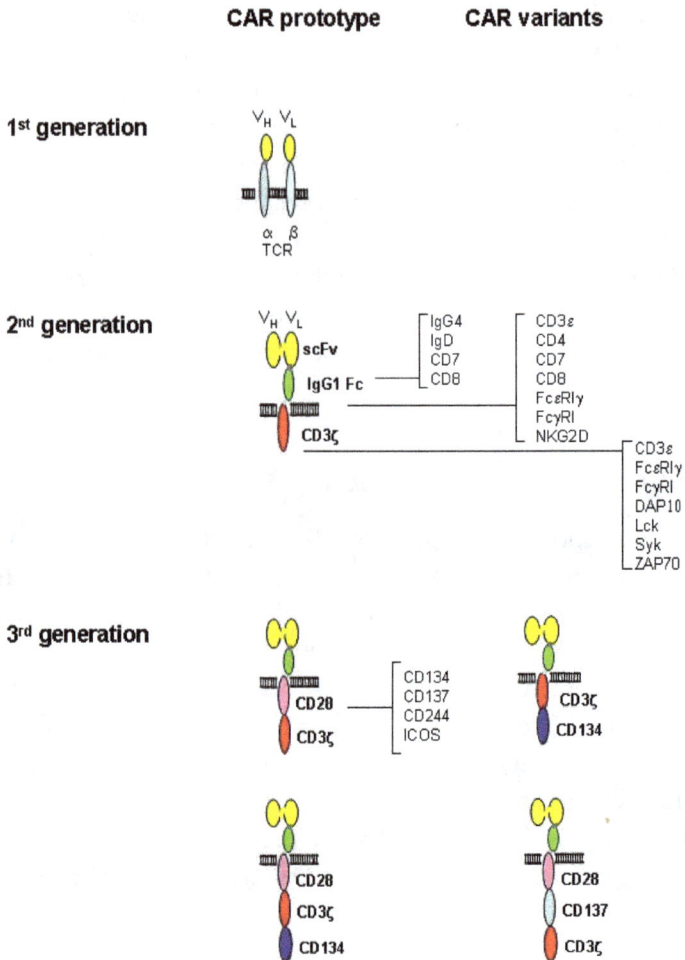

Figure 4.1. The "zoo" of chimeric antigen receptor configurations.

Chimeric antigen receptors of second and third generations are modularly composed of a single-chain antibody (scFv)-derived binding domain (yellow), an optional extracellular spacer (green), mostly the IgG1 Fc-hinge region, the trans-membrane, and signaling domain (red) which frequently is derived from CD3ζ but other intracellular signaling domains are likewise used. The CD3ζ domain is combined with one or two intracellular domains of the CD28 family in a third-generation CAR providing co-stimulatory signals.

Table 4.1. Specificity of CAR-modified T cells.

Antigen specificity	CAR format			Reference
	Spacer	TM Domain	Cytoplasmatic Domain	
CEA	—	CD3ζ	CD3ζ	10
CEA	IgG1	CD3ζ	CD3ζ	11, 12
CEA	IgG1	FcεRIγ	FcεRIγ	13, 14, 15, 12
CEA	CD8	FcεRIγ	FcεRIγ	16, 17
CEA	IgG1	FcεRIg	CD3ζ	14
CEA	CD8	CD28	CD28.CD3ζ	15
CEA	IgG1	CD28	CD28.CD3ζ	8, 18, 19
CEA	CD8Δ	CD3ζ	CD3ζ	20
CEA	—	CD3ε	CD3ε	20
CEA	IgG1	CD4	CD3ζ.CD134	21
CEA	IgG1	CD4	CD137.CD3ζ	21
NCAM	IgG1	CD3ζ	CD3ζ	11
NCAM	—	CD3ζ	CD3ζ	22
5T4	IgG1	CD3ζ	CD3ζ	11
fAchR	IgG1	CD3ζ	CD3ζ	23
EGP-2	CD8	FcεRIγ	FcεRIγ	24
CD19	IgG1	CD3ζ	CD3ζ	25
CD19	CD8	CD8	CD3ζ	26, 27
CD19	CD8	CD8	DAP10	26, 27
CD19	CD8	CD8	CD28.CD3ζ	26
CD19	IgG1	CD3ζ	CD3ζ.CD28	28
CD19	CD8	CD8	CD137.CD3ζ	26, 27
CD19	—	CD244	CD244.CD3ζ	29
CD20	IgG1	CD4	CD3ζ	30

(Continued)

Table 4.1. (*Continued*).

Antigen specificity	CAR format			Reference
	Spacer	TM Domain	Cytoplasmic Domain	
CD20	IgG4	CD4	CD28.137.CD3ζ	31
CD22	IgG1	CD4	CD3ζ	32
CD30	IgG1	CD3ζ	CD3ζ	33
CD33	IgG1	CD28	CD28.CD3ζ	34
CD33	IgG1	CD28	CD134.CD3ζ	34
CD33	IgG1	CD28	CD137.CD3ζ	34
CD33	IgG1	CD28	ICOS.CD3ζ	34
L1CAM	IgG1	CD4	CD3ζ	35, 36
HER2/neu	—	CD3ζ	CD3ζ	37
HER2/neu	IgG1	CD3ζ	CD3ζ	38
HER2/neu	IgG1	CD28	CD28.CD3ζ	38
HER2/neu	—	FcεRIγ	FcεRIγ	37
HER2/neu	CD8	CD28	CD28.CD3ζ	39
GD3	—	CD3ε	CD3ε	40
GD2	—	FcεRIγ	FcεRIγ	41
GD2	IgG1	CD28	CD28.CD3ζ	42
GD2	IgG1	CD134	CD134.CD3ζ	42
GD2	—	CD244	CD244.CD3ζ	29
GD2	IgG1	CD28	CD28.CD134.CD3ζ	42
VEGFR	CD8	CD8	CD3ζ	43
IL13Rα2	IgG4	CD4	CD3ζ	44, 45
IL13Ra2	IgG4	CD4	CD28.CD137.CD3ζ	44
NKG2DL	—	NKG2D	CD3ζ	46
Muc1	IgD	CD28	CD28.CD3ζ	47
Muc1	IgD	CD28	CD28.CD134.CD3ζ	47

(*Continued*)

Table 4.1. (*Continued*).

| Antigen specificity | CAR format | | | Reference |
	Spacer	TM Domain	Cytoplasmatic Domain	
Muc1	IgD	CD28	CD28.CD137.CD3ζ	47
Collagen type II	CD8	CD3ζ	CD3ζ	48
Collagen type II	—	FcεRIγ	FcεRIγ	49
PSMA	CD8	CD28	CD28.CD3ζ	50
PSCA	β2	CD3ζ	CD3ζ	51
HMW-MAA	—	FcεRIγ	FcεRIγ	52
Mesothelin	CD8	CD28	CD28.CD3ζ	53
Mesothelin	CD8	CD8	CD137.CD3ζ	53
Mesothelin	CD8	CD28	CD28.CD137.CD3ζ	53
HIV gp120	CD7	CD7	CD3ζ	54
HIV gp120	CD8	CD8	CD3ζ	54
HIV gp120	CD8	CD3ζ	CD3ζ	55
HBV S	IgG1	CD28	CD28.CD3ζ	56
HBV L	IgG1	CD28	CD28.CD3ζ	56
TNP	—	CD4	syk	57
TNP	—	CD8	syk	57
TNP	—	CD4	ZAP70	57
TNP	—	CD8	ZAP70	57
TNP	—	CD3ζ	CD3ζ	6
TNP	—	FcεRIγ	FcεRIγ	6

entirely composed of carbohydrates. Although the individual scFv domains mediate efficient binding of CAR-redirected T cells to these antigens, there are fundamental differences in the activation capacities. For instance, T cells redirected toward the high molecular weight-melanoma-associated antigen (HMW-MAA, MCSP) which is highly glycosylated are less efficiently activated than T cells redirected toward melanotransferrin, a polypeptide antigen, expressed on the same cell and bound with nearly same affinity (AH, HA, unpublished observations). The extent of cellular activation depends on the degree of receptor cross-linking[33] which seems to be dependent on the binding affinity and the targeted antigen itself.

The binding affinity has major impact on the efficacy in redirected T-cell activation. Compared to TCR–MHC interaction, the scFv domain of most CARs binds with high affinity, i.e., $k_D \leq 10^{-8}$ M. TCR binding with low affinities is thought to facilitate serial TCR triggering resulting in efficient T-cell activation even in the presence of small numbers of peptide-loaded MHCs per cell.[58] The CAR-targeted antigen, in contrast, is usually expressed in high numbers on the cell surface. CAR-mediated recognition of target cells is therefore likely to result in ligation of sufficient numbers of CAR molecules; the signaling threshold for cellular activation may be reached without the need for serial receptor triggering. Redirecting T cells via a series of CARs targeting the same epitope with different affinities revealed that the degree of receptor-mediated cellular activation is strictly correlated with scFv binding affinity when targeting solid-phase bound ligands.[38] The situation is different, however, when targeting the same ligands on the cell surface. Whereas low affinity binding induced less-efficient T-cell activation than high affinity binding to target cells, above a threshold, which was in that example at $k_D = 10^{-8}$ M, an increase in affinity did not result in an enhancement in T-cell activation.[38] There is obviously an upper limit in target cell-mediated T-cell activation which may be due to the spatial orientation of cell-to-cell interactions and the formation of antigen-engaged CAR synapses. CD28 co-stimulation does not alter that limit.[59]

Besides increasing affinity, binding to antigen can be improved by inserting a "spacer" domain between the scFv and transmembrane moiety of the CAR. Hinge and constant domains of the Ig family have shown to be good spacers. Although improving binding, this modification impairs, to some extent, CAR-mediated T-cell activation.[13]

In addition, the targeted epitope itself has a substantial impact on redirected T-cell activation. Using carcinoembryonic antigen (CEA) as an example, we revealed that upon targeting different epitopes on purified, immobilized antigen, the efficacy in CAR-mediated T-cell activation correlates with the binding affinity, irrespective of the targeted epitope.[60] However, in contrast, when targeting cell surface, CEA T cells are more efficiently activated when targeting the membrane proximal epitope than the membrane distal epitope, despite the latter binding with a higher affinity. The distal epitope when expressed in a more membrane proximal position, however, activated CAR-engineered T cells with higher efficiency than in the distal position indicating that not only the epitope itself but also the position of the targeted epitope has an influence on the efficiency of redirected T-cell activation.

The CAR expression level also influences the efficiency of T-cell activation.[61] Engineered T cells with high levels of CAR expression were activated by target cells with both high and low target antigen expression. In contrast, T cells with low-level CAR expression were only activated by target cells with high antigen expression.

Most antigens targeted by adoptive immunotherapy of malignant diseases are secreted or shedded by the respective tumor cells resulting in high concentrations of "tumor-associated antigens" in serum. These "soluble" antigens may block target cell recognition by binding to the scFv domain of CAR-engineered T cells. The situation may be circumvented by targeting an epitope on the membrane anchored part of the respective antigen which, however, will not be possible for the majority of targeted antigens. Using CEA as an example, we demonstrated that the anti-CEA CAR mediates specific and efficient T-cell activation by binding to CEA⁺ tumor cells even in the presence of CEA concentrations more than 10-fold higher than usually found in sera of tumor patients.[15] This may be due to the fact that the antigen in a membrane-bound form is present in preformed microdomains and thereby more likely generates CAR clustering and formation of a functional synapse than the same antigen in a mono or dimeric form in solution. Consequently, even low k_{on} values of the CAR-binding domain increase signaling over the threshold and drive

T-cell activation when encountering membrane-bound antigen. From the technical point of view, these observations suggest that an antigen-binding domain with low or intermediate binding affinities will be superior over high affinity binding domains since the latter may result in lower discrimination between the soluble and membrane antigen.

It is reasonable to assume that the clustering efficiency of CARs upon antigen engagement is moreover substantially influenced by the mobility of the targeted antigen in the cell membrane. This hypothesis is supported by the observation that decreasing the antigen fluidity within the membrane layer and increasing a cross-linkage of antigen by mild fixation results in an increase in CAR-mediated T-cell activation (A.A.H., H.A., unpublished data).

4.3.2 The extracellular "spacer" domain

Discrepancies in their T-cell activation capacities have revealed that some CARs are functionally active when the scFv is directly fused to the TM and signaling domain whereas others are not. CARs with CD3ζ signaling domain often require a "spacer" region between the antigen recognition and signaling domain. This requirement was first highlighted by Moritz and Groner.[62] In a report by Patel *et al.*,[54] CARs containing a CD7 or IgG1-derived spacer region demonstrated optimal target cell lysis, compared to CARs containing a CD8, truncated CD4, or truncated IgG1-derived spacer region. Although data suggested that extending the amino acid sequence between scFv and TM moiety is essential for optimal activity, the choice of CD7-, CD8-, or CD4-derived TM domains within these CARs may have also played a role in determining receptor efficiency. The spacer domain moreover increases the flexibility of the extracellular moiety and permits improved antigen binding by increasing the distance to the membrane. This is based on the observation that a recombinant CD3ζ-chain CAR with a CD8α hinge domain was more efficient in mediating T-cell activation than without.[63] While the IgG1 CH2CH3-hinge domain increased antigen binding and conjugate formation between effector and target cells, redirected T-cell activation was substantially impaired[13] indicating once again that improved antigen binding will not necessarily result in increased CAR signaling.

It was not until a comparative evaluation of CARs with and without spacer regions was published, that the issue was partly resolved.[11] Targeting antigens on the surface of tumor cells, the authors demonstrated that receptors containing scFv's specific for N-terminal epitopes did not require additional spacer regions whereas those targeting an antigen with C-terminal epitopes did. While an optimal distance seems to be required between T cell and target cell for T-cell activation, this hypothesis was disproved when a truncated antigen was targeted and optimal activity was still observed with CARs lacking a spacer region. The alternative hypothesis is that a spacer region is required for flexibility to target C-terminal epitopes on the cell membrane, which may have reduced accessibility to scFv binding.[11]

Recent work supports this earlier report. Targeting CD22[32] or CEA[64] revealed a higher degree of specific lysis when epitopes closer to the cell membrane were targeted. This may be explained by the flexibility model, i.e., increased flexibility in targeting membrane distal epitopes decreases the efficacy in CAR clustering; a kinetic-segregation model may be applied as well. This model, initially proposed by Davis and van der Merwe[65] and hypothesized to occur in CAR-engrafted T cell–target cell interactions, suggests that targeting membrane distal epitopes increases the size of the CAR–ligand pair, which in turn permits large phosphatase molecules such as CD45 to enter the synapse and negatively regulate TCR signaling.

Whereas an extracellular spacer domain is mandatory for stable expression of most CD3ζ signaling CARs on the surface of T cells,[62] a number of FcεRI γ-chain CARs can be successfully expressed in T cells without spacer domain.[13] Beside the TM domain, expression of CD3ζ-chain CARs is moreover modulated by interaction with components of preformed TCR synapse microdomains on the T-cell surface, a process which may be different for CD3ζ and FcεRI γ-chain CARs.

The choice of spacer region is still a matter of debate. The flexible immunoglobulin-like hinge region of CD8α was originally chosen[66] and has been used widely by other researchers since.[26,48,55,67] Efforts have been made to reduce preferential homo-dimerization, and thus increase functional CAR surface expression by mutating the cysteines in the CD8α hinge region.[20] Hinge regions from IgG subclasses have been most extensively used, especially IgG1[54,63] and to a lesser extent IgG4.[68] Although structurally very similar, with both containing two disulfide bonds, IgG4 is less susceptible to proteolytic cleavage than IgG1, which may make it more attractive than IgG1-derived spacers.[69] The inability of IgG4 to fix complement may also make it more appealing to prevent any unwanted effects on immune activity. More recently, IgD had been incorporated[47] as it has the longest and most flexible hinge region of any antibody, thus permitting antigen engagement in multiple orientations and at low antigen densities.[70] Problems with expression of IgD containing CARs were overcome via the incorporation of an additional IgG1 Fc in tandem with the IgD hinge.[71] This extremely large spacer region may not be amenable to all TAA targets but has proved efficient for targeting MUC-1.[71]

IgG1 spacer domains in the CAR extracellular domain have the as yet unrecognized, and potentially unwanted, effect of mediating binding to Fc receptors on innate immune cells resulting in "off-target" T-cell activation.[72] This situation can be overcome by introducing mutations into the IgG1 domain to prevent binding to Fc receptors yet without losing the capacity of the spacer to stabilize CD3ζ signaling CARs.

In summary, the design of CARs for targeting novel antigen must take spacer regions into account. With conflicting hypotheses being raised, it seems likely that the spacer/hinge regions may have multiple roles in mediating CAR flexibility and optimizing cell–cell distance. This difference may moreover depend on the particular antigen being targeted, and potentially, on the context in which the antigen is presented to the T cell, making testing of CARs with and without spacers to determine optimal activity essential.

4.3.3 *TM domain*

Cell surface receptors typically have a membrane spanning region consisting of 20–23 hydrophobic amino acids, rich in leucines, isoleucines, and valines. Early studies into CAR function utilized receptors with a variety of membrane spanning domains, including H2-Kb,[73] CD4,[37] CD7,[54] CD8,[54] FcεRIγ,[30] CD3ζ, and CD28. It is only recently that the role the TM domain has in mediating optimal CAR-mediated cellular activity has been demonstrated and a critical comparison has been undertaken. In one example, Heuser *et al.*[14] compared the TM and signaling domains of FcεRIγ and CD3ζ signaling CARs. Although it is the signaling, rather than the TM domain which affected surface expression in primary human T cells, it was also demonstrated that T cells harboring CD3ζ-based receptors performed better when they contained the FcεRIγ compared with the native CD3ζ TM domain. This difference in function was not seen when the cytoplasmic domain was derived from the FcεRI γ-chain. The intricacies of the CD3ζ TM domain itself have shown to have subtle but important consequences for CAR activity. CD3ζ dimerization is a prerequisite for optimal activity.[74] The charged aspartic acid in the TM domain also regulates antigen sensitivity by permitting incorporation of the CD3ζ-based CAR into the endogenous TCR complex.[75]

One interesting aspect of the FcεRIγ TM domain is its ability to mediate hetero-dimerization with endogenous CD3ζ,[57] a feature that has the potential to enable a truncated CAR lacking the cytoplasmic domain to mediate T-cell activation indirectly via the endogenous CD3ζ.[76] We have demonstrated that this signaling in trans also occurs when using the CD3ζ TM domain as discussed below.

In a more basic evaluation of different TM domains on CD3ζ-mediated mast-cell degranulation, Gosse *et al.*[74] engrafted CARs with the plasma membrane spanning regions of IL2Rα, CD71, PTPα, or CD45. Optimum activity was observed with the CD3ζ TM domain containing receptor. Interestingly, mast cell degranulation, calcium flux, and tyrosine phosphorylation of the signaling domains was dependent on CAR incorporation into lipid-rafts following cross-linking, which is dependent upon the TM domain.

The CD4 TM domain, first published in 1998,[30] has been extensively used in preclinical studies and clinical trials by the Jensen group (City of Hope National Medical Center, USA). Whereas CARs with the highly truncated CD4 TM domain and the IgG hinge region as spacer form homodimers, CARs with the CD4 TM domain alone, and no extracellular spacer, failed to form dimers.[57] The CD8 extra-cellular hinge region has been extensively used due to its flexible structure and multiple cysteine residues which provide rigid dimeri-zation. As such, several groups have extended the CD8 hinge to also encompass the CD8 TM domain.[26,76] Alternatively, NKG2D was investigated for use in CARs by fusing the complete NKG2D extracellular, TM, and cytoplasmic domain to the CD3ζ cytoplasmic domain.[46]

In summary, there is little rationale behind the choice of TM domain in the reports discussed above, with different groups having their "favorite" TM domain. Although downstream effects on recep-tor function are evident, there has, so far, been no critical comparison published on the most commonly used membrane spanning regions in CARs making such a comparative evaluation overdue.

4.3.4 The signaling domain

The CAR signaling domain couples antigen recognition to the intra-cellular signaling pathway. It was first shown by Kuwana et al.[77] and subsequently by Gross et al.[1] that the scFv-binding domain could be fused to the TCR constant domain and the receptor generated could induce redirected T-cell activation. As research progressed, the signal-ing domains from CD3ζ or FcεRIγ were incorporated into the CAR[6] permitting the choice of signaling domain rather than relying on the TCR signaling components. The FcεRI γ-chain and the CD3 ζ-chain since then have been widely used due to the three immunoreceptor tyrosine activation motifs (ITAM) in the CD3 ζ-chain and one highly homologous ITAM in the γ-chain which become phosphorylated upon CAR engagement of antigen. A large body of work highlight the importance and sufficiency of CD3ζ alone for efficient T-cell acti-vation[78,79]; FcR γ-chain receptors, however, are used due to the claims that the first CD3ζ ITAM can mediate T-cell anergy.[80,81]

Efforts to compare γ- and ζ-chain signaling CARs have met with conflicting results. CARs with γ-chain are more stably expressed in T cells than ζ-chain CARs,[33] whereas the latter are more stably expressed in non-T cell effectors, e.g., NK cells and neutrophils, than

FcR γ-chain CARs.[82] The different requirements for stable expression may be due to different interactions with the endogenous CD3/TCR complex in engineered T cells. Heuser *et al.*[14] found that γ-chain CARs had a higher activation threshold than ζ-chain CARs. Although T-cell responses were equal over short-term, ζ-chain CARs outperformed γ-chain CARs in long-term co-culture experiments. Moreover, γ-chain CARs were inhibited more by soluble antigen than CD3ζ receptors. This *in vitro* observation is supported by *in vivo* work in which murine T cells with ζ-chain CAR had superior anti-tumor responses to γ-chain CAR-engrafted T cells.[17] However, Ren-Heidenreich *et al.*[24] were unable to identify any overall differences in cytolytic activity, cytokine secretion, and activation-induced cell death (AICD) between cells expressing FcεRIγ or CD3ζ CARs.

It appears that the CD3ζ chain in CARs can be significantly reduced in size without a comparable loss of function due to the regulatory role of CD3ζ in T-cell activation contrasting with the widely held notion of a purely stimulatory role.[83] This idea is elegantly demonstrated by Chae *et al.*[76] who used a CAR-like molecule to demonstrate that CD3ζ signaling components containing Tyr>Phe mutant in the third ITAM were largely tolerated; however, mutations to the first or second ITAM had a significant negative impact on T-cell activation. Interestingly, the first ITAM alone could recapitulate much of the function of the wild-type receptor with regards to cytokine secretion and induction of AICD. This work has been supported by other reports which indicate that CD3ζ has an ordered system of ITAM phosphorylation on antigen stimulation with tyrosines in the first ITAM required for phosphorylation of subsequent tyrosines in ITAM 2 and 3.[84] This may be a rationale for the development of CARs with a smaller signaling domain without a concomitant loss of function.

CAR with ζ- and γ-chain are likely to be integrated into the CD3/TCR complex with different efficiencies. In TCR[low] T-cell clones, ζ-chain CARs enhanced expression of the endogeneous CD3 and TCR$\alpha\beta$ which was not observed in T cells engineered with γ-chain CARs (A.A.H., H.A., unpublished data) implying that ζ-chain CARs restore and stabilize TCR expression by interaction whereas γ-chain CARs do not. This is in line with the observation that signaling can be mediated by cytoplasmic-deficient CARs via hetero-dimerization with the TCR signaling machinery[74] (J.B., D.E.G., unpublished data). There are substantial differences in the activation capacities of γ- and ζ-chain CARs. T cells engineered with ζ-chain CAR were more efficiently activated *in vivo* than with γ-chain CAR[80] which was reported for NK cells and neutrophils as well.[82]

Alternatively, other signaling moieties were also successfully exploited for use in CAR-redirected T-cell activation, e.g., the CD3 ε-chain.[20] CD3ε could activate engrafted T cells equivalently to CD3ζ-CAR expressing cells on immobilized antigen; however, when presented in the context of cell membrane antigen, the CD3ε-CAR failed to engage its cognate ligand. One advantage of CD3ε signaling is that these CARs would preferentially hetero-dimerize with CD3δ and CD3γ which would not only increase the number of functional CAR molecules on the cell surface, but also give a more rounded T-cell response as there would be a more diverse array of ITAMs involved compared to CD3ζ-based CARs.

TCR signaling is frequently impaired in tumor-bearing individuals[85] which provides the rationale to design receptors with a TCR downstream signaling kinase like Syk or ZAP70.[57,85–87] Interestingly, Syk-based receptors were more efficient in T-cell activation than ZAP70-based receptors. Incorporating additional components into the basic scFv.CD3ζ format has been used to improve receptor function. One approach is to combine signaling domains with signaling-intermediate molecules which endow the CAR with kinase activity, e.g., by fusing the Src-kinase Lck to the C-terminus of CD3ζ to improve CAR function.[73] The combined use of co-receptors such as CD4 would seem a reasonable approach to optimize CD3ζ phosphorylation following antigen engagement since TCR signaling requires CD4 or CD8 cooperation to recruit kinases such as Lck to the synapse.[88] Whether CD4 integration into a CAR would lead to a more realistic signaling cascade is questionable and there would be various safety concerns regarding the potential for non-specific CAR activity.

Accessory molecules can also modulate CAR-mediated T-cell activation. For instance, cytolysis by a γ-chain CAR is substantially co-modulated by ICAM-1 (CD54) on the target cell.[61] CD3ζ-redirected

T cells are similarly induced by ICAM-1 on the target cell to release IL-2, although in low quantities.[89] Furthermore, endogenous CD2 can act as a co-stimulator to induce IL2 secretion from CAR-engrafted T cells.[90] As a consequence, T cells redirected by immuno-receptors are co-modulated in certain, but not always predictable ways by accessory molecules expressed on tumor cells. To modulate CAR-triggered T-cell activation more precisely and independently of ligands on the surface of tumor cells, a co-stimulatory signal needs to be simultaneously present with the CD3ζ signal upon antigen engagement. This requirement resulted in the development of the third-generation CARs with combined CD3ζ and CD28 signaling domains.

4.3.5 Third-generation CARs: Co-stimulation combined with CD3ζ- or FcRγ-driven activation

According to the dual signal model, a co-stimulatory signal in addition to the primary TCR/CD3 signal is required for full T-cell activation and prevention from AICD or anergy. CD3ζ signaling without CD28 co-stimulation is not sufficient to initiate redirected activation of resting T cells as shown in a transgenic mouse model.[91,92] For cancer immunotherapy, the ability to include a co-stimulatory domain in the CAR offers a great advantage. The major advantage is that co-stimulation tunes T-cell effector functions by suppressing inhibitory and/or strengthening stimulatory signals. Co-stimulation is physiologically provided by CD28 family members, dependent on the stage of T-cell activation, in particular by CD28 upon primary activation and by 4–1BB (CD137) or OX40 (CD134) to sustain T-cell activation.

CD28 co-stimulation is usually delivered by antigen presenting cells (APCs) but most therapeutically targeted cells including tumor cells do not express the stimulatory ligands B7.1 (CD80) or B7.2 (CD86). To overcome this limitation, Alvarez-Vallina *et al.* generated a T-cell clone expressing both a CD3ζ and CD28 signaling CAR which enhance cellular activation.[93] Technical limitations in co-expression of two CARs in peripheral T cells were overcome by the design of a CAR with fused signaling domains of CD3ζ and CD28 in the intracellular moiety.[8,89,94]

In completely activated T cells, CD28 co-stimulation modulates CAR-triggered effector functions in a specific manner. CD28-CD3ζ CAR-redirected T cells are efficiently protected from AICD and exhibit increased proliferation and IFN-γ secretion, whereas antigen-redirected cytolysis is not altered by CD28 co-stimulation.[8,90,95] Accordingly, Beecham *et al.* demonstrated that CD28 co-stimulation is required for prolonged polyclonal expansion of CAR-engineered T cells.[96] Increase in T-cell survival and amplification *in vivo* results in a prolonged redirected T-cell response and in improved target cell lysis making third-generation CAR favorable for clinical use.

CD28 co-stimulation is moreover indispensable to induce IL-2 secretion by redirected T cells[90] which use IL-2 in an autocrine manner to increase their amplification. Accordingly, T cells engineered with combined CD3ζ-CD28 signaling CARs secrete substantial amounts of IL-2 upon antigen binding without exogenous B7/CD28 co-stimulation. Notably, specific cytolysis by engineered T cells is not affected by CD28 co-stimulation.[90] IL-2 secreted in high concentrations by engineered T cells into the microenvironment upon redirected activation may attract in turn a second wave of inflammatory cells, thus locally enhancing the anti-tumor effect. Regulatory suppressor T cells (Treg cells), on the other hand, require IL-2 for survival since they are itself not capable of IL-2 production. IL-2 secreted by redirected T cells upon activation thereby supports Treg survival in a paracrine manner resulting in increased repression of the T-body's response.[97] This effect can be eliminated by modification of the CD28 lck binding domain in order to selectively repress IL-2 secretion without altering the co-stimulatory activity of CD28.

CD28 co-stimulation, on the other hand, has substantial impact on preventing T-cell repression and induction of anergy. Whereas IL-10 does not have an effect on redirected T-cell activation *in vitro*, TGF-β1 represses CD3ζ-CAR-mediated T-cell proliferation, but not IFN-γ secretion and redirected cytolysis. TGF-β1 mediated repression can be counteracted by CD28 co-stimulation through combined CD28-CD3ζ signaling CAR.[18]

Other members of the CD28 co-stimulatory family can substitute some but not all effector functions mediated by CD28 co-stimulation

in CAR-redirected T cells.[21] Although there is some functional overlap, each member of the CD28 family has distinct functions, depending on the stimuli and the antigenic history of the lymphocytes.
Physiologically, OX40 (CD134) and 4–1BB (CD137) are expressed
after CD28 signaling and are thought to be involved in prolonging
the immune response and in generating T-cell memory. In late T-cell
activation, OX40–OX40L interactions prolong IL-2 secretion and
trigger the generation of memory T cells. The improved anti-tumor
capacity of CARs harboring CD137 domains may be related to pro-
survival signals.[21,26,98] Moreover, CD137 signaling increases TCR-
induced proliferation and cytokine production as well as CTL
generation and tumor rejection. CD137 signaling in combination
with CD3ζ is superior to CD28ζ or CD28-CD137-CD3ζ CARs in
anti-leukemic activity *in vivo*.[98,99] A thorough comparison of combined signaling CARs with either CD28, CD134, or CD137 co-
stimulatory domain, each in series with CD3ζ, in CD4+ and CD8+ T
cells revealed that T-cell proliferation is dramatically increased by
CD28, but not by CD134 or CD137 co-stimulation, which is
observed for both CD4+ and CD8+ T cells.[21,100] Noteworthy, IL-2
secretion is only induced upon CD28, but not upon CD134 or
CD137 co-stimulation, whereas IFN-γ secretion is increased by each
CD28, CD134, and CD137 co-stimulation. Cytotoxicity, however, is
not significantly altered by CD134 or CD137 co-stimulation compared to a moderate improvement upon CD28 co-stimulation. In
CD4+ T cells, AICD is diminished upon co-stimulation compared to
CD3ζ signaling only, whereas in CD8+ T cells, CD137 co-stimulation, but not CD28 and CD134 co-stimulation, prevents AICD.
Taken together, each co-stimulus modulates a distinct pattern of
T-cell effector functions in its own manner although there is some
overlap. In resting T cells, CD28, ICOS, and CD134, each combined
in series with CD3ζ into an immunoreceptor, enhanced receptor triggered, antigen-specific cytolytic activities and ICOS, CD134, and
CD137, respectively, confers self-sufficient clonal expansion upon
antigen encounter and enhanced cytokine secretion.[34]

Whether addition of non-CD28 receptor family domains could
provide a distinct activation profile is an area that has received little

attention. CD2 can provide co-stimulation to first-generation CAR-engrafted T cells and its ligand, CD58, is often downregulated on a number of tumors making this an interesting alternative.[90] Another CD2 family receptor, 2B4 (CD244), has also been tested in T cells and NK cells and has proven efficient in modulating T-cell effector function.[90]

From a structural viewpoint, it is interesting to note that the specific order of domains within the cytoplasmic part of CAR appears to be important for optimal activity. Optimal IL-2 production was seen when the CD28 domain was membrane proximal and the CD3ζ domain membrane distal in comparison to a CD3ζ-CD28 orientation.[34,95] The same holds for CD137.[98] CD134, however, requires the distal position and appears to be inactive in the proximal position.[95] This may be explained by the requirement of CD28 and CD137 to be in close proximity to the membrane to recruit interacting partners, e.g., PI3K. This is plausible as membrane lipids are substrates of the PI3K-Akt pathway. Alternatively, or in addition, the relevant TM domain derived from the co-stimulatory moiety may position the CAR in the correct lipid membrane compartments required for co-stimulation; the latter explanation is supported by recent evidence suggesting that specific CD28 signaling motifs are required to position CD28 in the correct cell membrane regions for co-stimulatory function.[101]

Other signaling domains, such as those of p56lck and CD4, have also been combined with CD3ζ to improve CAR-mediated activation. Particularly, CD4-CD3ζ, CD3ζ-lck, and CD28-CD3ζ-lck signaling enhanced ZAP70 phosphorylation and IL-2 secretion and decreased the signaling threshold of receptor-engrafted T cells.[102] Because these combined signaling immunoreceptors have been analyzed so far only in established T-cell lines, their impact on primary T cells with respect to antigen-triggered cytolysis, proliferation, and cytokine secretion has still to be investigated.

What is clear from fused signaling constructs is that optimal positioning of the two primary signal and co-stimulatory elements is essential for optimal CAR function. The combined signaling approach provides full T-cell activation by signal 1 and signal 2 upon engagement of the targeted antigen. Separate expression of the individual signaling elements, i.e., CD3ζ and CD28, is a strategy that could be reinvestigated having the advantage that different scFv's could be fused to each signaling moiety thus providing dual tumor antigen specificity.

4.3.6 CARs with combined co-stimulatory domains

While the CD3ζ signaling domain can be combined individually with the CD28, CD134, or CD137 co-stimulatory domain into one polypeptide chain, it is a small step to then combine co-stimulatory domains along with CD3ζ in a single co-signaling moiety. Since CD137 co-stimulation prevents AICD and, on the other hand, CD28 sustains T-cell proliferation and cytokine secretion, including IL-2, a combination of both the CD28 and CD137 domain along with CD3ζ may be advantageous for sustaining T-cell survival and proliferation *in vivo*. CD137 was initially compared with CD28, CD134, and ICOS in tandem with CD3ζ.[34] This created the first receptor incorporating two co-stimulatory domains.[42] The combination of CD28, CD137, and CD3ζ can provide superior antigen-specific cell expansion and IL-2 secretion as was demonstrated by another study comparing CD28ζ with CD28.CD137.ζ CARs.[31] June and coworkers most recently reported that combined CD28 and CD137 signaling together with CD3ζ is superior in sustaining persistence and superior anti-leukemic activity of redirected T cells *in vivo*.[53,99] On the other hand, one-polypeptide chain immunoreceptors with triple signaling domains may inefficiently recruit the signaling molecules that trigger the downstream phosphorylation cascades. As a consequence, triple-signaling receptors may be unstable in the T-cell membrane. Taken this into account, it remains to be elucidated whether triple signaling CARs in general are superior in triggering persistence and function of redirected T cells.

4.4 Clinical Studies

As shown in Table 4.1, many CARs specific to tumor-associated antigens have been generated so far covering a wide spectrum of cancer targets. A few of these have already been applied in phase-I clinical trials using T-bodies for cancer therapy, and several others are already ongoing, or in an advanced phase of preparation[81,103] (Table 4.2). Only a few of these studies have been published and the rest of the information in this section is derived from registries and from personal communications.

Cell Genesys conducted phase I clinical trials in colorectal patients using the anti-TAG72-ζ CAR made from the humanized CC49 mAb.[104] This trial, however, was terminated due to the identification of anti-idiotypic antibodies in the patient sera, which caused difficulty in interpretation of the results. The group of Junghans tested 24 doses of CEA-specific CAR-bearing lymphocytes,[105] with a total dose of up to 10^{11} cells per patient. The treatment was reported to be adequately tolerated, with only two minor adverse effects observed in two colorectal carcinoma patients. Further studies by the same group are ongoing CEA and PSMA T-bodies with CD28 co-stimulatory CARs.[106] The NIH trial targeting CEA[+] carcinoma by CD28-ζ CAR T cells, however, was terminated due to severe gastrointestinal and pulmonary toxicities.[107] Hwu and colleagues[108] at the NCI conducted a phase I clinical trial in ovarian cancer patients using T-bodies expressing a ζ CAR specific to human α-folate receptor, also known as FBP. This trial demonstrated that large numbers of gene-modified tumor-reactive T cells can be safely given to patients, but these cells do not persist in large numbers in the long term. No reduction in tumor burden was seen in any patient. Tracking ^{111}In-labeled adoptively transferred T cells revealed that the T cells did not localize to the tumor, except in one patient where some signal was detected in a peritoneal deposit. PCR analysis showed that gene-modified T cells were present in the circulation in large numbers for the first 2 days after transfer, but these quickly declined and became barely detectable 1 month later in most patients. Five out of eight patients who received a dose escalation of T cells in combination with high-dose IL-2 experienced some grade 3 to 4 treatment-related toxicity that was probably due to IL-2 administration, which could be managed using standard measures. Patients in cohort 2 who received T cells with dual specificity (reactive with both FR and allogeneic cells) followed by

Table 4.2. Clinical trials using CAR-modified T cells.

Target antigen	Disease	Chimeric antigen receptor endodomain	Gene transfer	ClinicalTrial. gov identifier	Center
CD19	CLL	CD28-CD3ζ	RV	NCT00466531	MSKCC
CD19	B-ALL	CD28-CD3ζ	RV	NCT01044069	MSKCC
CD19	Leukemia	CD28-CD3ζ	RV	NCT01416974	MSKCC
CD19	Leukemia/lymphoma	CD28-CD3ζ	RV	NCT00924326	NCI
CD19	Leukemia/lymphoma	CD28-CD3ζ	RV	NCT01087294	NCI
CD19	Leukemia/lymphoma	CD28-CD3ζ vs CD3ζ	RV	NCT00586391	BCM
CD19	B-NHL/CLL	CD28-CD3ζ vs CD3ζ	RV	NCT00608270	BCM
CD19	Advanced B-NHL/CLL	CD28-CD3ζ vs CD3ζ	RV	NCT00709033	BCM
CD19	ALL	CD3ζ	RV	NCT01195480	UCL
CD19	ALL post-HSCT	CD28-CD3ζ	RV	NCT00840853	BCM
CD19	Leukemia/lymphoma	CD137-CD3ζ vs CD3ζ	LV	NCT00891215	UP
CD19	B-lymphoid malignancies Post-HSCT	CD28-CD3ζ	EP	NCT00968760	MDACC
CD19	B-lineage lymphoid Malignancies post-UCBT	CD28-CD3ζ	EP	NCT01362452	MDACC
CD19	B-NHL	CD3ζ	RV	NCT01493453	CHMAN
CD19	B-NHL	CD3ζ	LV	NCT01318317	COH

(Continued)

Table 4.2.   (*Continued*).

Target antigen	Disease	Chimeric antigen receptor endodomain	Gene transfer	ClinicalTrial. gov identifier	Center
CD20	Mantle cell lymphoma/ Indolent B-NHL	CD28-CD137-CD3ζ	EP	NCT00621452	FHCRC
PMSA	Prostate carcinoma	CD28-CD3ζ	RV	NCT01140373	MSKCC
PMSA	Prostate carcinoma	CD3ζ	RV	NCT00664196	RWMC
CEA	Breast cancer	CD28-CD3ζ	RV	NCT00673829	RWMC
CEA	Colorectal cancer	CD28-CD3ζ	RV	NCT00673322	RWMC
Her2/neu	Lung cancer	CD28-CD3ζ	RV	NCT00889954	BCM
Her2/neu	Osteosarcoma	CD28-CD3ζ	RV	NCT00902044	BCM
Her2/neu	Glioblastoma	CD28-CD3ζ	RV	NCT01109095	BCM
GD2	Neuroblastoma	CD3ζ	RV	NCT00085930	BCM
Ig k light chain	B-NHL and B-CLL	CD28-CD3ζ vs CD3ζ	RV	NCT00881920	BCM
FBP	Ovarian cancer	γ	RV	NCT00019136	NCI
TAG-72	Colorectal cancer	CD3ζ	RV	McArthur	CellGenesys
Carbo-anhyd IX	Renal cancer	γ	RV	Gratama	DDHC
IL-13 receptor	Glioblastoma	CD3ζ	EP	Jensen	COH
CD171	Neuroblastoma	CD3ζ	EP	Jensen	COH

(*Continued*)

Table 4.2. (*Continued*).

Target antigen	Disease	Chimeric antigen receptor endodomain	Gene transfer	ClinicalTrial. gov identifier	Center
CEA	Colorectal cancer	CD3ζ	RV	Hawkins	Manchester
CEA	Prostate cancer	CD3ζ	RV	Hawkins	Manchester
Mesothelin	Pancreatic cancer	CD3ζ	RV	June	UP
Lewis-Y	Myeloma	CD3ζ	RV	Kershaw	Melbourne
CD20	Lymphoma	CD28-CD3ζ	RV	Cooper	MDACC

MSKCC, Memorial Sloan Kettering Cancer Center; NCI, National Cancer Institute; BCM, Baylor College of Medicine; RWMC, Roger Williams Medical Center; UCL, University College London; UP, University of Pennsylvania; MDACC, M.D. Anderson Cancer Center; CHMAN, Christie's Hospital, University of Manchester; COH, City of Hope Medical Center; FHCRC, Fred Hutchinson Cancer Research Center; DDHC, Daniel den Hoed Cancer Center Rotterdam; EP, Electroporation.

immunization with allogeneic peripheral blood mononuclear cells, experienced relatively mild side effects with grade 1 to 2 symptoms. Neutralizing antibodies were found in some of the patient sera, specific to the murine anti-FBP MoV18 mAb.

The Rotterdam's ATTACK Partners reported a phase I clinical trial in renal cell cancer (RCC), using autologous T lymphocytes modified with a CAR specific for carboxy anhydrase IX.[109,110] Infusions of the T-bodies into patients were initially well tolerated. However, after four to five infusions, all three patients began to develop liver enzyme abnormalities. This was explained by the reactivity of the genetically modified cells with low levels of carboxy anhydrase IX expressed on the bile duct epithelium, limiting treatment to only low doses of CAR-expressing T-bodies. The results of this study showed that the engineered T cells exert CAR-directed functions *in vivo*. Several patients in the trial also developed antibodies to the murine G250 scFv. Because of these side effects, this trial was put on hold and awaits renewal pending the application of systemic anti-carboxy anhydrase IX antibodies to block antigen expression on the bile duct.[111]

The results of a safety/feasibility trial using human CTL clones redirected at metastatic neuroblastoma was reported by the group of Jenssen in the City of Hope.[36] In this trial, CD8$^+$ CTL clones were transfected with anti-CD171 CAR and the selection of suicide expression enzyme HyTK. Six children with recurrent/refractory neuroblastoma received 12 infusions. No overt toxicities to tissues known to express the CD171 adhesion molecule were observed. The persistence of the modified CTL in the circulation was short (1–7 days) in patients with bulky disease, but significantly longer (42 days) in a patient with limited disease burden. The authors suggest this pilot study set the stage for clinical trial in the context of minimal residual disease. This group also developed an efficient procedure to manufacture large numbers of patient's-derived T cells.[112]

Most of the currently ongoing trials have been designed based on lessons learned from the preclinical animal models, e.g., use of humanized scFv, inclusion of the CD28 co-stimulatory domain, use of lentiviral vectors for T-cell modification, inclusion of homeostatic interleukins in the *ex vivo* procedures used to prepare the T-bodies, transfection of both CD4⁺ and CD8⁺ T cells, or use of T-bodies made from autologous T cells that are also specific to viral antigens such as EBV and influenza. Very importantly, in several of these studies, lymphoablative pretreatments are being used to precondition the patients before the administration of T-bodies. Several of these trials target blood-borne tumors such as lymphoma or leukemia, using anti-CD19, anti-CD20, or anti-CD22 scFv's from humanized antibodies. Although less challenging than solid tumors, the results of using T-bodies against these targets will demonstrate certain efficacy and thereby paving the way for applying the T-body approach against more challenging solid tumors. A proof-of-concept clinical trial in which patients with relapsed or refractory indolent B-cell lymphoma or mantle cell lymphoma were treated with autologous T cells genetically modified by electroporation with a vector plasmid encoding a CD20-specific chimeric TCR and neomycin resistance gene was reported by the Fred Hutchinson Cancer Research Center.[113] Transfected cells were immunophenotypically similar to CD8⁺ effector cells and showed CD20-specific cytotoxicity *in vitro*. Seven patients received a total of 20 T-cell infusions, with minimal toxicities. Modified T cells persisted *in vivo* 1 to 3 weeks in the first 3 patients, who received T cells produced by limiting dilution methods, but persisted 5 to 9 weeks in the next 4 patients who received T cells produced in bulk cultures followed by 14 days of low-dose subcutaneous IL-2 injections. Of the 7 treated patients, 2 maintained a previous complete response, 1 achieved a partial response, and 4 had stable disease. Another recent clinical trial by the group of Brenner[114] have shown a correlation between the persistence of the adoptive transferred T-bodies and the clinical outcome. The group in Baylor College of Medicine engineered Epstein–Barr virus-specific CTLs to express a CAR directed to the disialoganglioside GD2, a non-viral tumor-associated antigen expressed by human neuroblastoma cells, aiming that the redirected CTLs would receive optimal co-stimulation after engagement of their native receptors, enhancing survival and anti-tumor activity mediated through their chimeric receptors. When administered to individuals with neuroblastoma, the EBV-specific CTLs expressing a chimeric GD2-specific receptor indeed survive longer than T cells activated by anti-CD3 antibody and expressing the same chimeric receptor but lacking virus specificity.

In a most recent trial, CD28-4-1BB-ζ CAR-modified T cells redirected toward CD19 produced complete and lasting remission of refractory CD19$^+$ B-cell chronic lymphocytic leukemia (CLL).[115,116] T cells were effective even at low dosage levels, i.e., about 1.5 ′ 10^5 cells per kg. CAR T cells expanded *in vivo* more than 1,000-fold compared to the initial engraftment level and persisted in the peripheral blood and bone marrow for at least 6 months. T-cell infusions had no acute toxic effects; the only adverse event noted was a grade-3 tumor lysis syndrome in one patient. Interestingly, there was a delayed increase in the serum levels of pro-inflammatory cytokines which paralleled the clinical symptoms and coincided with the elimination of leukemia cells from the bone marrow. Infusion of these genetically modified cells seemed safe and was associated with spectacular tumor regression in those subjects tested. In contrast to this trial, T cells modified with a CD28-ζ CAR targeting the same antigen rapidly disappeared from circulation, although some responses were observed.[117] The prolonged persistence of CD28–4-1BB-ζ CAR-modified T cells is probably due to the cooperative effect of CD28 and 4-1BB in sustaining T-cell survival. However, anti-CD19 CAR T cells may be continuously re-stimulated by emerging CD19$^+$ early B-cell progenitors and thus maintain a pool of activated CAR T cells in the long term, which may additionally contribute to sustain tumor control.

Despite recent success, two fatal serious adverse events occurred after infusion of T cells with co-stimulatory CARs. In the NIH trial, the events were attributed to the "on-target off-organ" activation of engineered T cells due to the ubiquitous target expression on healthy tissues.[118] The adverse event seen after treatment of a CLL patient with CD28-4-1BB CAR T cells was attributed to an extravasation of a latent bacterial infection after subsequent lymphodepletion.[119]

Despite recent severe adverse events, adoptive cell therapy with CAR T cells showed safe, feasible, and with some anti-tumor activity. Introduction of genes to T cells using retroviral or lentiviral vectors has been proved safe and it is clear today that the risk of leukemia that occurred in patients receiving retroviral vector-mediated gene transfer into hematopoietic stem cells does not exist for mature T cells[120] although rarely observed in an experimental setting.[121] Potential severe side effects seen in some patients in the clinical trials reported above are manageable, side effects due to "off-organ" T-cell activation need to be carefully ruled out. In line with that, a careful selection of the antibody whose scFv will serve to redirect the T-bodies, both in terms of specificity and affinity, will diminish the risk of damaging essential healthy tissues. Side effects of IL-2 could be controlled and hopefully will be prevented when the persistence of the engineered T cells in the body will be improved.

4.5 Perspectives

The two obvious advantages of the CAR strategy are the production of immune cells with antibody defined specificity suitable for therapeutic application and the MHC-independent targeting of a variety of cell surface proteins. The antibody-derived binding domain allows to target a variety of antigens assuming a scFv antibody is available. These antigens include polypeptide antigens as well as carbohydrates, including CA19–9, Muc-1, HMW-MAA, and others. Tumor cells as well as infected cells frequently downregulate proteins associated with antigen processing and presentation including MHC. This mechanism effectively renders the tumor cell invisible to T-cell attack. Consequently, the direct, MHC-independent targeting of cell surface proteins directly through the CAR antibody-derived binding domain avoids the need for antigen presentation on MHC molecules and thereby make the tumor vulnerable to T-cell attack. Tumor cells which do not express the targeted antigen, however, are not attacked by redirected T cells making tumor cell variants with loss of antigen expression, a major source of potential tumor relapse after initially successful treatment.

There is a major mechanistic advantage of CARs as compared to recombinant TCR-based approaches. The potential risk of generating novel or autoimmune T-cell specificities through receptor mis-pairing is highly reduced compared to heterologous TCRs (see Chapter 3). CARs use the TCR downstream signaling pathway for T-cell activation but do not physically pair with TCR $\alpha\beta$ chains as do recombinant $\alpha\beta$ TCRs to a substantial extent. Current data available imply that CD3ζ signaling CARs can physically interact with CD3ζ in the TCR synapse[75] but this does not impact TCR specificity and T-cell activation capacity upon antigen engagement. CAR-engineered T cells can thereby be stimulated in a specific manner via the TCR and the CAR as shown for CAR-modified, EBV-specific T cells[122] and others.

Due to the modular design, CAR molecules can be relatively easily modified by replacing the binding or signaling domain or to include additional signaling domains thereby modulating or strengthening the potency of CAR signaling. Powerful selection systems, such as phage display, provide a plethora of binding domains to target virtually any cell surface protein. By mutagenesis, the affinity of a given scFv can be increased or reduced.[38,123] The high affinity for antigen binding provided by the scFv as compared to that of the TCR suggests that engineered T cells may have improved antigen specificity.

However, there are potential balances to these advantages which need careful consideration. The high complexity of the recognition and signaling process implies that the universal optimized configuration of a CAR for each antigen may not exist. This is likely to be due to the extreme diversity of structures which can be targeted by CARs in contrast to the much more standardized TCR recognition and activation process through MHC-presented antigen. The high affinity for antigen may moreover be deleterious to CAR-redirected T-cell function. On the other hand, TCR-like scFvs specific for MHC peptide have been developed[123] which could be employed to generate TCR-like CARs; however, the advantage of these MHC-restricted CARs over the use of recombinant TCRs remains unclear aside from the fact that the CAR consists of a single expressed protein while the TCR approach requires co-expression of both α and β TCR chains.

Can CAR-engineered T cells efficiently recycle lytic capacity to kill multiple targets? There is *in vitro* evidence suggesting that T cells can do so, although formal *in vivo* confirmation is still lacking. Indeed, understanding the mechanisms of CAR-mediated killing of target cells is still at an early stage. Recent studies suggest that CARs containing a CD3ζ domain can interact with the endogenous TCR/CD3 complex and that CARs lacking functional ITAMs can still activate the transduced T cell suggesting cis-signaling mediated by CAR interaction with TCR signaling receptors[75] (J.B., unpublished observations).

As discussed above, the CAR structure plays key roles in optimizing the function of engineered T cells; however, critical to the success of therapy will be the choice of targeted antigen and the relationship of the targeted epitope of that particular antigen with respect to the topography at the plasma membrane of the target cell. Data suggest that an optimal T-cell-to-target cell spacing distance is required. As such, there may be no one-fits-all receptor and empirical testing of receptors for the specific antigen may be required to identify the best for that particular antigen.

Furthermore, the high affinity binding domain on the surface of T-bodies may affect trafficking and homing to sites of tumor. For example, does strong binding to target antigen cause the T cells to be locally trapped? What effect would secreted antigens such as CEA have upon *in vivo* trafficking of engineered T cells? Soluble CEA does not appear to impact upon targeted T-cell function *in vitro* but the impact upon T-cell trafficking remains unknown. Most recent observations point out that T cells with a low to medium affinity CAR target to CEA⁺ tumors even in presence of increased levels of serum CEA (Chmielewski, A.A.H., H.A., unpublished data). Moreover, there is evidence indicating that scFv binding affinity is important in terms of target cell killing while equally, the level of antigen expression is also important. In essence, low-affinity scFv can direct the activity of engineered T cells only against targets with abundant levels of antigen while high-affinity scFv are effective against low or high antigen levels on target cells. Consequently, whether the affinity of the scFv potentially impacts T-cell trafficking in the tumor-bearing patient is also currently unclear.

Homing and migration of engineered T cells may also be manipulated through expression of chemokine receptors[124] without or alongside CARs.[125] Alternatively, isolated T-cell subpopulations which express, or lack, certain chemokine receptors may be used for tumor targeting. These modifications together could be an approach to more fully control the *in vivo* targeting of engineered T cells. Studies investigating these issues in animal models are now gaining momentum and are likely to provide important insights in the near future.

A beneficial effector cell-to-target cell ratio at the tumor site is likely to be required for efficient target cell lysis. An estimation based on clinical data, however, is not yet available although higher numbers of engineered T cells, i.e., $> 10^{10}$ cells per dose, are likely to increase efficacy. Adoptive transfer of a low dose of 10^7 CAR T cells produced an anti-tumor response in a most recent trial.[116] Rapid *ex vivo* expansion to large numbers of engineered T cells is feasible, the protocols currently used to transduce and amplify the engineered T cells, however, may not generate cells with the optimal phenotype for adoptive transfer. Consideration of this point is made elsewhere (Chapter 2).

The design of third-generation CAR integrating CD3ζ signaling with co-stimulatory pathways, however, still raises a number of questions. How the various co-stimulatory signals are qualitatively and quantitatively integrated in the T cell? How can the activity of various co-stimulatory signals be optimized for redirected elimination of tumor cells without negatively affecting other bystander tissues? Different types of co-stimulation may result in different patterns of cellular activation properties. At present, the inclusion of a co-stimulatory signaling domain, such as CD28 or 4-1BB, within the CAR appears to result in improved therapeutic activity in certain model systems. However, these constructs generate multiple signals using a single binding event thereby avoiding many of the cellular checkpoints employed by T cells to avoid uncontrolled proliferation. Consequently, determining the long-term safety profile of these receptors will be paramount. Therefore, several co-stimulatory pathways and their combinations are worthy of exploration for use in the immunoreceptor strategy in the near future. How can co-stimulation prevent receptor-grafted T cells from becoming anergic? Does stimulation through ICAM-1 complement CD28 co-stimulation *in vivo*? Can an OX40 co-stimulatory CAR contribute to the induction of tumor-specific memory? Co-stimulation and subsequent cytokine release may moreover be used for co-opting by-stander immune responses, for instance activation of tumor-infiltrating T and NK cells, thereby improving the anti-tumor immune response.

As far as safety concerns, autoimmunity may result from "on-target off-organ" T-cell activation, such as targeting healthy tissues which physiologically express low levels of the targeted antigen, as has been the case in CAR targeting carboxy anhydrase IX (G250), a renal cell carcinoma antigen also expressed at low levels on bile duct epithelia.[109,110] In this case, toxicity was controlled using steroids to deplete modified T cells. Since there are a very few genuine tumor-specific antigens, the choice of antigen for the immunoreceptor needs to be carefully considered and control methods adopted to ensure off-target toxicities are kept to a minimum. Indeed, in the case of T-cell adoptive therapy of melanoma, off-target toxicities generally predict for anti-tumor responses.[126] However, controlling the engineered T cell *in vivo* represents an important option. Novel gene suicide systems using tagged receptor molecules which can be targeted by T-cell-depleting antibodies *in vivo*[127] have recently been developed to permit specific depletion of the engineered T cells rather than the need for total T-cell depletion strategies. Recently, the use of zinc-finger nuclease technology has permitted successful editing of the TCR in CAR-engrafted T cells. Such technology has the potential to generate allogeneic T-cell pools for transfer to multiple donors without the risk of graft versus host responses.[128]

While CARs use the TCR signaling machinery, the T-body strategy is obviously not restricted to redirect T cells; monocytes, macrophages as well as NK cells can be specifically redirected by ζ-signaling CARs as well.[129,130] Whether redirected non-T cells have a benefit in tumor elimination needs to be explored in appropriate *in vivo* models.

Taken together, strong arguments support the principle and development of this form of immunotherapy. Since the initial descriptions of the approach nearly 20 years ago, most of the early technological hurdles concerning the methods to generate these cells have been largely overcome and an increasing focus of the field is upon understanding the mechanisms of immune-directed therapies *in vivo*, how effective CAR-redirected T cells are when compared to "natural T cells," and how to further improve the approach in order to ensure that the methodology can become more widely available. Despite a number of unresolved questions, engineered T cells redirected by third- or fourth-generation CAR, combined with ablation of suppressor cells, provides an attractive strategy to implement an effective tumor-specific T-cell response. Clinical trials that are implemented and currently expanded will furthermore determine whether redirected T-cell activation by CARs can re-establish immune surveillance of tumor cells.

4.6 Summary

The CAR (immunoreceptor, T-body) is a recombinant one-polypeptide chain receptor molecule consisting of an antibody-derived binding domain and a "spacer" in the extracellular moiety, a trans-membrane and a signaling domain with or without co-stimulatory elements in the intracellular moiety. Due to antibody-mediated binding, CARs circumvent MHC restriction in recognition, target a variety of antigens of different structures, and initiate strong T-cell activation upon antigen binding. Since the initial descriptions of the approach nearly 20 years ago, most of the early technological hurdles to generate CARs have been largely overcome and the structural requirements of a CAR become increasingly clear. Each and every component of the CAR design offers some functional property for optimization which must be taken into consideration when designing novel receptors. As an example, the extracellular domain appears to be the structural component most likely to affect specificity, and also T-cell activation through its target cell-binding properties such as affinity. As such, the benefit of extracellular spacers between recognition and TM moiety must be tested for each cognate antigen epitope under investigation. TM domains also appear to significantly impact on optimal CAR activity that can be related to surface expression and interactions with other membrane molecules, including the TCR signaling machinery, which may or may not impinge on CAR activity. Finally, the optimization of the cytoplasmic signaling domain, without or with one or two co-stimulatory moieties, has to take into account the extracellular and TM domains. Whereas CARs were so far optimized for function in T cells, their design for optimized activity in other types of effector immune cells like NK cells, macrophages, or neutrophils, needs to be re-evaluated. Despite a number of unresolved questions, engineered T cells redirected by third- or fourth-generation CAR, combined with ablation of suppressor cells, provides an attractive strategy to implement an effective tumor-specific T-cell response. An increasing focus of the field is upon understanding the mechanisms of redirected therapies *in vivo*, how effective T-bodies are when compared to TCR-modified and "natural T cells", and how to further improve the approach in order to ensure that the methodology can become more economic and widely available. Clinical trials that are implemented and currently expanded will furthermore determine whether redirected T-cell activation by CARs can control tumor growth.

References

1. Gross, G., Waks, T., and Eshhar, Z. (1989). Expression of immunoglobulin-T cell receptor chimeric molecules as functional receptors with antibody-type specificity. *Proc Natl Acad Sci USA* 86, 10024–10028.

2. Goverman, J., Gomez, S.M., Segesman, K.D., Hunkapiller, Y., Lang, W.E., and Hood, L. (1990). Chimeric immunoglobulin — Tcell receptor proteins form functional receptors: Implications for T cell receptor complex formation and activation. *Cell* 60, 929–939.

3. Gross, G. and Eshhar, Z. (1992). Endowing T cells with antibody specificity using chimeric T cell receptors. *Faseb J* 6, 3370–3378.

4. Huston, J.S., Levinson, D., Mudget-Hunter, M., Tai, M.S., Novotny, J., Margolies, M.N., Ridge, R.J., Bruccoleri, R.E., Haber, E., Crea, R., and Oppermann, H. (1988). *Proc Natl Acad Sci USA* 85, 5879–5884.

5. Bird, R.E., Hardman, K.D., Jacobson, J.W., Johnson, S., Kaufman, B.M., Lee, S.M., Lee, T., Pope, S.H., Riordan, G.S., and Withlow, M. (1988). *Science* 242, 423–426.

6. Eshhar, Z., Waks, T., Gross, G., and Schindler, D.G. (1993). Specific activation and targeting of cytotoxic lymphocytes through chimeric single chains consisting of antibody-binding domains and the gamma or zeta subunits of the immunoglobulin and T cell receptors. *Proc Natl Acad Sci USA* 90, 720–724.

7. Finney, H.M., Lawson, A.D., Bebbington, C.R., and Weir, A.N. (1998). Chimeric receptors providing both primary and co-stimulatory signaling in T cells from a single gene product. *J Immunol* 161, 2791–2797.

8. Hombach, A., Wieczarkowiecz, A., Marquardt, T., Heuser, C., Usai, L., Pohl, C., Seliger, B., and Abken, H. (2001). Tumor specific T cell activation by recombinant immunoreceptors: CD3zeta signaling and CD28 costimulation are simultaneously required for efficient IL-2 secretion and can be integrated into one combined CD28/CD3zeta signaling receptor molecule. *J Immunol* 167, 6123–6131.

9. Finney, H.M., Akbar, A.N., and Lawson, A.D. (2004). Activation of resting human primary T cells with chimeric receptors: Co-stimulation from CD28, inducible costimulator, CD134, and CD137 in series with signals from the TCR zeta chain. *J Immunol* 172, 104–113.

10. Gilham, D.E., O'Neil, A., Hughes, C., Guest, R.D., Kirillova, N., Lehane, M., *et al.* (2002). Primary polyclonal human T lymphocytes targeted to carcinoembryonic antigens and neural cell adhesion molecule tumor antigens by CD3zeta-based chimeric immune receptors. *J Immunother* 25, 139–151.

11. Guest, R.D., Hawkins, R.E., Kirillova, N., Cheadle, E.J., Arnold, J., O'Neill, A., Irlam, J., Chester, K.A., Kemshead, J.T., Shaw, D.M., Embleton, M.J., Stern, P.L., and Gilham, D.E. (2005). The role of extracellular spacer regions in the optimal design of chimeric immune receptors: Evaluation of four different scFvs and antigens. *J Immunother* 28, 203–211.

12. Hombach, A., Schneider, C., Sent, D., Koch, D., Willemsen, R.A., Diehl, V., *et al.* (2000). An entirely humanized CD3 zeta-chain signaling receptor that directs peripheral blood T cells to specific lysis of carcinoembryonic antigen-positive tumor cells. *Int J Cancer* 88, 115–120.

13. Hombach, A., Heuser, C., Gerken, M., Fischer, B., Lewalter, K., Diehl, V., Pohl, C., and Abken, H. (2000). T cell activation by recombinant FcεRI γ-chain immune receptors: An extracellular spacer domain impairs antigen dependent T cell activation but not antigen recognition. *Gene Ther* 7, 1067–1075.

14. Heuser, C., Hombach, A., Lösch, C., Manista, K., Sircar, R., and Abken, H. (2003). T-cell activation by recombinant immunoreceptors: Impact of the intracellular signalling domain on the stability of receptor expression and antigen-specific activation of grafted T cells. *Gene Ther* 10, 1408–1419.

15. Hombach, A., Koch, D., Sircar, R., Heuser, C., Diehl, V., Kruis, W., Pohl, C., and Abken, H. (1999). A chimeric receptor that selectively targets membrane-bound carcinoembryonic antigen (mCEA) in presence of soluble CEA. *Gene Ther* 6, 300–304.

16. Darcy, P.K., Haynes, N.M., Snook, M.B., Trapani, J.A., Cerruti, L., Jane, S.M., *et al.* (2000). Redirected perforin-dependent lysis of colon carcinoma by *ex vivo* genetically engineered CTL. *J Immunol* 164, 3705–3712.

17. Haynes, N.M., Snook, M.B., Trapani, J.A., Cerruti, L., Jane, S.M., Smyth, M.J., and Darcy, P.K. (2001). Redirecting mouse CTL against colon carcinoma: Superior signaling efficacy of single-chain variable domain chimeras containing TCR-zeta vs Fc epsilon RI-gamma. *J Immunol* 166, 182–187.

18. Koehler, H., Kofler, D., Hombach, A., and Abken, H. (2007). CD28 costimulation overcomes TGF-β mediated repression of proliferation of redirected human CD4+ and CD8+ T-cells in an anti-tumor cell attack. *Cancer Res* 67, 2265–2273.

19. Hombach, A., Schlimper, C., Sievers, E., Frank, S., Schild, H.H., Sauerbruch, T., *et al.* (2006). A recombinant anti-carcinoembryonic antigen immunoreceptor with combined CD3zeta-CD28 signalling targets T cells from colorectal cancer patients against their tumour cells. *Gut* 55, 1156–1164.

20. Nolan, K.F., Yun, C.O., Akamatsu, Y., Murphy, J.C., Leung, S.O., Beecham, E.J., and Junghans, R.P. (1999). Bypassing immunization: Optimized design of "designer T cells" against carcinoembryonic antigen (CEA)-expressing tumors, and lack of suppression by soluble CEA. *Clin Cancer Res* 5, 3928–3941.

21. Hombach, A. and Abken, H. (2007). Costimulation tunes tumor-specific activation of redirected T cells in adoptive immunotherapy. *Cancer Immunol Immunother* 56, 731–737.

22. Sheen, A.J., Sherlock, D.J., Irlam, J., Hawkins, R.E., and Gilham, D.E. (2003). T lymphocytes isolated from patients with advanced colorectal cancer are suitable for gene immunotherapy approaches. *Br J Cancer* 88, 1119–1127.

23. Gattenlohner, S., Marx, A., Markfort, B., Pscherer, S., Landmeier, S, Juergens, H., *et al.* (2006). Rhabdomyosarcoma lysis by T cells expressing a human autoantibody-based chimeric receptor targeting the fetal acetylcholine receptor. *Cancer Res* 66, 24–28.

24. Ren-Heidenreich, L., Mordini, R., Hayman, G.T., Siebenlist, R., and LeFever, A. (2002). Comparison of the TCR zeta-chain with the FcR gamma-chain in chimeric TCR constructs for T cell activation and apoptosis. *Cancer Immunol Immunother* 51, 417–423.

25. Landmeier, S., Altvater, B., Pscherer, S., Eing, B.R., Kuehn, J., Rooney, C.M., *et al.* (2007). Gene-engineered varicella-zoster virus reactive CD4+ cytotoxic T cells exert tumor-specific effector function. *Cancer Res* 67, 8335–8343.

26. Marin, V., Kakuda, H., Dander, E., Imai, C., Campana, D., Biondi, A., and D'Amico, G. (2007). Enhancement of the anti-leukemic activity of cytokine induced killer cells with an anti-CD19 chimeric receptor delivering a 4–1BB-zeta activating signal. *Exp Hematol* 35, 1388–1397.

27. Imai, C., Iwamoto, S., and Campana, D. (2005). Genetic modification of primary natural killer cells overcomes inhibitory signals and induces specific killing of leukemic cells. *Blood* 106, 376–83.

28. Loskog, A., Giandomenico, V., Rossig, C., Pule, M., Dotti, G., and Brenner, M.K. (2006). Addition of the CD28 signaling domain to chimeric T-cell receptors enhances chimeric T-cell resistance to T regulatory cells. *Leukemia* 20, 1819–1828.

29. Altvater, B., Landmeier, S., Pscherer, S., Temme, J., Juergens, H., Pule, M., *et al.* (2009). 2B4 (CD244) signaling via chimeric receptors costimulates tumor-antigen specific proliferation and *in vitro* expansion of human T cells. *Cancer Immunol Immunother* 58, 1991–2001.

30. Jensen, M., Tan, G., Forman, S., Wu, A.M., and Raubitschek, A. (1998). CD20 is a molecular target for scFvFc:zeta receptor redirected T cells: Implications for cellular immunotherapy of CD20+ malignancy. *Biol Blood Marrow Transplant* 4, 75–83.

31. Wang, J., Jensen, M., Lin, Y., Sui, X., Chen, E., Lindgren, C.G., Till, B., Raubitschek, A., Forman, S.J., Qian, X., James, S., Greenberg, P., Riddell, S., and Press, O.W. (2007). Optimizing adoptive polyclonal T cell immunotherapy of lymphomas, using a chimeric T cell receptor possessing CD28 and CD137 costimulatory domains. *Hum Gene Ther* 18, 712–725.

32. James, S.E., Greenberg, P.D., Jensen, M.C., Lin, Y., Wang, J., Till, B.G., Raubitschek, A.A., Forman, S.J., and Press, O.W. (2008). Antigen sensitivity of CD22-specific chimeric TCR is modulated by target epitope distance from the cell membrane. *J Immunol* 180, 7028–7038.

33. Hombach, A., Heuser, C., Sircar, R., Tillmann, T., Diehl, V., Pohl, C., and Abken, H. (1998). An anti-CD30 chimeric receptor that mediates CD3-ζ independent T-cell activation against Hodgkin's lymphoma cells in the presence of soluble CD30. *Cancer Res* 58, 1116–1119.

34. Finney, H.M., Akbar, A.N., and Lawson, A.D. (2004). Activation of resting human primary T cells with chimeric receptors: Costimulation from CD28, inducible costimulator, CD134, and CD137 in series with signals from the TCR zeta chain. *J Immunol* 172, 104–113.

35. Gonzalez, S., Naranjo, A., Serrano, L.M., Chang, W.C., Wright, C.L., and Jensen, M.C. (2004). Genetic engineering of cytolytic T lymphocytes for adoptive T-cell therapy of neuroblastoma. *J Gene Med* 6, 704–711.

36. Park, J.R., Digiusto, D.L., Slovak, M., Wright, C., Naranjo, A., Wagner, J., Meechoovet, H.B., Bautista, C., Chang, W.C., Ostberg, J.R., and Jensen, M.C. (2007). Adoptive transfer of chimeric antigen receptor re-directed cytolytic T lymphocyte clones in patients with neuroblastoma. *Mol Ther* 15, 825–833.

37. Stancovski, I., Schindler, D.G., Waks, T., Yarden, Y., Sela, M., and Eshhar, Z. (1993). Targeting of T lymphocytes to Neu/HER2-expressing cells using chimeric single chain Fv receptors. *J Immunol* 151, 6577–6582.

38. Chmielewski, M., Hombach, A., Heuser, C., Adams, G.P., and Abken, H. (2004). T cell activation by antibody-like immunoreceptors: Increase in affinity of the scFv domain above threshold does not increase T cell activation against antigen-positive target cells but decreases selectivity. *J Immunol* 173, 7647–7653.

39. Moeller, M., Haynes, N.M., Trapani, J.A., Teng, M.W., Jackson, J.T., Tanner, J.E., *et al.* (2004). A functional role for CD28 costimulation in tumor recognition by single-chain receptor-modified T cells. *Cancer Gene Ther* 11, 371–379.

40. Yun, C.O., Nolan, K.F., Beecham, E.J., Reisfeld, R.A., and Junghans, R.P. (2000). Targeting of T lymphocytes to melanoma cells through chimeric anti-GD3 immunoglobulin T-cell receptors. *Neoplasia* 2, 449–459.

41. Rossig, C., Bollard, C.M., Nuchtern, J.G., Merchant, D.A., and Brenner, M.K. (2001). Targeting of G(D2)-positive tumor cells by human T lymphocytes engineered to express chimeric T-cell receptor genes. *Int J Cancer* 94, 228–236.

42. Pule, M.A., Straathof, K.C., Dotti, G., Heslop, H.E., Rooney, C.M., and Brenner, M.K. (2005). A chimeric T cell antigen receptor that augments cytokine release and supports clonal expansion of primary human T cells. *Mol Ther* 12, 933–941.

43. Niederman, T.M., Ghogawala, Z., Carter, B.S., Tompkins, H.S., Russell, M.M., and Mulligan, R.C. (2000). Antitumor activity of cytotoxic T lymphocytes engineered to target vascular endothelial growth factor receptors. *Proc Natl Acad Sci USA* 99, 7009–7714.

44. Chang, L., Chang, W.C., McNamara, G., Aguilar, B., Ostberg, J.R., and Jensen, M.C. (2007). Transgene-enforced co-stimulation of CD4+ T cells leads to enhanced and sustained anti-tumor effector functioning. *Cytotherapy* 9, 771–784.

45. Kahlon, K.S., Brown, C., Cooper, L.J., Raubitschek, A., Forman, S.J., and Jensen, M.C. (2004). Specific recognition and killing of glioblastoma multiforme by interleukin 13-zetakine redirected cytolytic T cells. *Cancer Res* 64, 9160–9166.

46. Zhang, T., Barber, A., and Sentman, C.L. (2006). Generation of antitumor responses by genetic modification of primary human T cells with a chimeric NKG2D receptor. *Cancer Res* 66, 5927–5933.

47. Wilkie, S., Picco, G., Foster, J., Davies, D.M., Julien, S., Cooper, L., Arif, S., Mather, S.J., Taylor-Papadimitriou, J., Burchell, J.M., and Maher, J. (2008). Retargeting of human T cells to tumor-associated MUC1: The evolution of a chimeric antigen receptor. *J Immunol* 180, 4901–4909.

48. Annenkov, A. and Chernajovsky, Y. (2000). Engineering mouse T lymphocytes specific to type II collagen by transduction with a chimeric receptor consisting of a single chain Fv and TCR zeta. *Gene Ther* 7, 714–722.

49. Annenkov, A.E., Moyes, S.P., Eshhar, Z., Mageed, R.A. and Chernajovsky, Y. (1998). Loss of original antigenic specificity in T cell hybridomas transduced with a chimeric receptor containing single-chain Fv of an anti-collagen antibody and Fc epsilonRI-signaling gamma subunit. *J Immunol* 161, 6604–6613.

50. Maher, J., Brentjens, R.J., Gunset, G., Riviere, I., and Sadelain, M. (2002). Human T-lymphocyte cytotoxicity and proliferation directed by a single chimeric TCRzeta/CD28 receptor. *Nat Biotechnol* 20, 70–75.

51. Morgenroth, A., Cartellieri, M., Schmitz, M., Günes, S., Weigle, B., Bachmann, M., Abken, H., Rieber, E.P., and Temme, A. (2007). Targeting of tumor cells expressing the prostate stem cell antigen (PSCA) using genetically engineered T-cells. *Prostate* 67, 1121–1131.

52. Reinhold, U., Liu, L., Lüdtke-Handjery, H.C., Heuser, C., Hombach, A., Wang, X., Tilgen, W., Ferrone, S., and Abken, H. (1999). Specific lysis of melanoma cells by receptor grafted T cells is enhanced by anti-idiotypic monoclonal antibodies directed to the scFv domain of the receptor. *J Invest Dermatol* 112, 744–750.

53. Carpenito, C., Milone, M.C., Hassan, R., Simonet, J.C., Lakhal, M., Suhoski, M.M., Varela-Rohena, A., Haines, K.M., Heitjan, D.F., Albelda, S.M., Carroll, R.G., Riley, J.L., Pastan, I., and June, C.H. (2009). Control of large, established tumor xenografts with genetically retargeted human T cells containing CD28 and CD137 domains. *Proc Natl Acad Sci USA* 106, 3360–3365.

54. Patel, S.D., Moskalenko, M., Smith, D., Maske, B., Finer, M.H., and McArthur, J.G. (1999). Impact of chimeric immune receptor extracellular protein domains on T cell function. *Gene Ther* 6, 412–419.

55. Masiero, S., Del Vecchio, C., Gavioli, R., Mattiuzzo, G., Cusi, M.G., Micheli, L., Gennari, F., Siccardi, A., Marasco, W.A., Palù, G., and Parolin, C. (2005). T-cell engineering by a chimeric T-cell receptor with antibody-type specificity for the HIV-1 gp120. *Gene Ther* 12, 299–310.

56. Bohne, F., Chmielewski, M., Ebert, G., Wiegmann, K., Kürschner, T., Schulze, A., Urban, S., Krönke, M., Abken, H., and Protzer, U. (2008). T cells redirected against hepatitis B virus surface proteins eliminate infected hepatocytes. *Gastroenterol* 134, 239–247.

57. Fitzer-Attas, C.J., Schindler, D.G., Waks, T., and Eshhar, Z. (1998). Harnessing Syk family tyrosine kinases as signaling domains for chimeric single chain of the variable domain receptors: Optimal design for T cell activation. *J Immunol* 160, 145–154.

58. Valitutti, S. and Lanzavecchia, A. (1997). Serial triggering of TCRs: A basis for the sensitivity and specificity of antigen recognition. *Immunol Today* 18, 299–304.

59. Chmielewski, M., Hombach, A.A., and Abken, H. (2011). CD28 cosignalling does not affect the activation threshold in a chimeric antigen receptor-redirected T-cell attack. *Gene Ther* 18, 62–72.

60. Hombach, A., Schildgen, V., Heuser, C., Finnern, R., Gilham, D., and Abken, H. (2007). T cell activation by antibody-like immunoreceptors: The position of the binding epitope within the target molecule determines the efficiency of activation of redirected T cells. *J Immunol* 178, 4650–4657.

61. Weijtens, M.E., Hart, E.H., and Bolhuis, R.L. (2000). Functional balance between T cell chimeric receptor density and tumor associated antigen density: CTL mediated cytolysis and lymphokine production. *Gene Ther* 7, 35–42.

62. Moritz, D. and Groner, B. (1995). A spacer region between the single chain antibody- and the CD3 zeta-chain domain of chimeric T cell receptor components is required for efficient ligand binding and signaling activity. *Gene Ther* 2, 539–546.

63. Ren-Heidenreich, L., Hayman, G.T., and Trevor, K.T. (2000). Specific targeting of EGP-2+ tumor cells by primary lymphocytes modified with chimeric T cell receptors. *Hum Gene Ther* 11, 9–19.

64. Hombach, A.A., Schildgen, V., Heuser, C., Finnern, R., Gilham, D., Abken, H. (2007). T cell activation by antibody-like immunoreceptors: The position of the binding epitope within the target molecule determines the efficiency of activation of redirected T cells. *J Immunol* 178, 4650–4657.

65. Davis, S.J. and van der Merwe, P.A. (2006). The kinetic-segregation model: TCR triggering and beyond. *Nat Immunol* 7, 803–809.

66. Moritz, D., Wels, W., Mattern, J., and Groner, B. (1994). Cytotoxic T lymphocytes with a grafted recognition specificity for ERBB2-expressing tumor cells. *Proc Natl Acad Sci USA* 91, 4318–4322.

67. Shibaguchi, H., Luo, N.X., Kuroki, M., Zhao, J., Huang, J., Hachimine, K., Kinugasa, T., and Kuroki, M. (2006). A fully human chimeric immune receptor for retargeting T-cells to CEA-expressing tumor cells. *Anticancer Res* 26, 4067–4072.

68. Kradin, R., Kurnick, J., Gifford, J., Pinto, C., Preffer, F., and Lazarus, D. (1989). Adoptive immunotherapy with interleukin-2 (IL-2) results in diminished IL-2 production by stimulated peripheral blood lymphocytes. *J Clin Immunol* 9, 378–385.

69. Solomon, A., Gramse, M., and Havemann, K. (1978). Proteolytic cleavage of human IgG molecules by neutral proteases of polymorphonuclear leukocytes. *Eur J Immunol* 8, 782–785.

70. Loset, G.A., Roux, K.H., Zhu, P., Michaelsen, T.E., and Sandlie, I. (2004). Differential segmental flexibility and reach dictate the antigen binding mode of chimeric IgD and IgM: Implications for the function of the B cell receptor. *J Immunol* 172, 2925–2934.

71. Maher, J. and Wilkie, S. (2009). CAR mechanics: Driving T cells into the MUC of cancer. *Cancer Res* 69, 4559–4562.

72. Hombach, A., Hombach, A.A., and Abken, H. (2010). Adoptive immunotherapy with genetically engineered T cells: Modification of the IgG1 Fc 'spacer' domain in the extracellular moiety of chimeric antigen receptors avoids 'off-target' activation and unintended initiation of an innate immune response. *Gene Ther* 17, 1206–1213.

73. Geiger, T.L., Nguyen, P., Leitenberg, D., and Flavell, R.A. (2001). Integrated src kinase and costimulatory activity enhances signal transduction through single-chain chimeric receptors in T lymphocytes. *Blood* 98, 2364–2371.

74. Gosse, J.A., Wagenknecht-Wiesner, A., Holowka, D., and Baird, B. (2005). Transmembrane sequences are determinants of immunoreceptor signaling. *J Immunol* 175, 2123–2131.

75. Bridgeman, J.S., Hawkins, R.E., Bagley, S., Blaylock, M., Holland, M., and Gilham, D.E. (2010). The optimal antigen response of chimeric antigen receptors harboring the CD3zeta transmembrane domain is dependent upon incorporation of the receptor into the endogenous TCR/CD3 complex. *J Immunol* 184, 6938–6949.

76. Chae, W.J., Lee, H.K., Han, J.H., Kim, S.W., Bothwell, A.L., Morio, T., and Lee, S.K. (2004). Qualitatively differential regulation of T cell activation and apoptosis by T cell receptor zeta chain ITAMs and their tyrosine residues. *Int Immunol* 16, 1225–1236.

77. Kuwana, Y., Asakura, Y., Utsunomiya, N., Nakanishi, M., Arata, Y., Itoh, S., Nagase, F., and Kurosawa, Y. (1987). Expression of chimeric receptor composed of immunoglobulin-derived V regions and T-cell receptor-derived C regions. *Biochem Biophys Res Commun* 149, 960–968.

78. Irving, B.A. and Weiss, A. (1991). The cytoplasmic domain of the T cell receptor zeta chain is sufficient to couple to receptor-associated signal transduction pathways. *Cell* 64, 891–901.

79. Geiger, T.L., Leitenberg, D., and Flavell, R.A. (1999). The TCR zeta-chain immunoreceptor tyrosine-based activation motifs are sufficient for the activation and differentiation of primary T lymphocytes. *J Immunol* 162, 5931–5939.

80. Kersh, E.N., Kersh, G.J., and Allen, P.M. (1999). Partially phosphorylated T cell receptor zeta molecules can inhibit T cell activation. *J Exp Med* 190, 1627–1636.

81. Eshhar, Z. (2008). The T-body approach: Redirecting T cells with antibody specificity. *Handb Exp Pharmacol* 181, 329–342.

82. Roberts, M.R., Cooke, K.S., Tran, A.C., Smith, K.A., Lin, W.Y., Wang, M., Dull, T.J., Farson, D., Zsebo, K.M., and Finer, M.H. (1998). Antigen-specific cytolysis by neutrophils and NK cells expressing chimeric immune receptors bearing zeta or gamma signaling domains. *J Immunol* 161, 375–384.

83. Pitcher, L.A. and van Oers, N.S. (2003). T-cell receptor signal transmission: Who gives an ITAM? *Trends Immunol* 24, 554–560.

84. Kersh, E.N., Shaw, A.S., and Allen, P.M. (1998). Fidelity of T cell activation through multistep T cell receptor zeta phosphorylation. *Science* 281, 572–575.

85. Mizoguchi, H., O'Shea, J.J., Longo, D.L., Loeffler, C.M., McVicar, D.W., and Ochoa, A.C. (1992). Alterations in signal transduction molecules in T lymphocytes from tumor-bearing mice. *Science* 258, 1795–1798.

86. Eshhar, Z. and Fitzer-Attas, C.J. (1998). Tyrosine kinase chimeras for antigen-selective T-body therapy. *Adv Drug Deliv Rev* 31, 171–182.

87. Fitzer-Attas, C.J., Schindler, D.G., Waks, T. and Eshhar, Z. (1997). Direct T cell activation by chimeric single chain Fv-Syk promotes Syk-Cbl association and Cbl phosphorylation. *J Biol Chem* 272, 8551–8557.

88. Ledbetter, J.A., Deans, J.P., Aruffo, A., Grosmaire, L.S., Kanner, S.B., Bolen, J.B., and Schieven, G.L. (1993). CD4, CD8 and the role of CD45 in T-cell activation. *Curr Opin Immunol* 5, 334–340.

89. Hombach, A., Sent, D., Schneider, C., Heuser, C., Koch, D., Pohl, C., Seliger, B., and Abken, H. (2001). T cell activation by recombinant receptors: CD28 costimulation is required for IL-2 secretion and receptor mediated T cell proliferation but does not affect receptor mediated target cell lysis. *Cancer Res* 61, 1976–1982.

90. Cheadle, E.J., Rothwell, D.G., Bridgeman, J.S., Sheard, V.E., Hawkins, R.E., and Gilham, D.E. (2011). Ligation of the CD2 co-stimulatory receptor enhances IL-2 production from first-generation chimeric antigen receptor T cells. *Gene Ther* Dec 1. doi: 10.1038/gt.2011.192 [Epub ahead of print].

91. Brocker, T. and Karjalainen, K. (1995). Signals through T cell receptor-zeta chain alone are insufficient to prime resting T lymphocytes. *J Exp Med* 181, 1653–1659.

92. Brocker, T. (2000). Chimeric Fv-zeta or Fv-epsilon receptors are not sufficient to induce activation or cytokine production in peripheral T cells. *Blood* 96, 1999–2001.

93. Alvarez-Vallina, L. and Hawkins, R.E. (1996). Antigen-specific targeting of CD28-mediated T cell co-stimulation using chimeric single-chain antibody variable fragment-CD28 receptors. *Eur J Immunol* 26, 2304–2309.

94. Finney, H.M., Lawson, A.D., Bebbington, C.R., and Weir, A.N. (1998). Chimeric receptors providing both primary and costimulatory signaling in T cells from a single gene product. *J Immunol* 161, 2791–2797.

95. Abken, H., Hombach, A., Heuser, C., Kronfeld, K., and Seliger, B. (2002). Tuning tumor-specific T cell activation: A matter of costimulation? *Trends Immunol* 23, 240–245.

96. Beecham, E.J., Ma, Q., Ripley, R., and Junghans, R.P. (2000). Coupling CD28 co-stimulation to immunoglobulin T-cell receptor molecules: The dynamics of T-cell proliferation and death. *J Immunother* 23, 631–642.

97. Kofler, D.M., Chmielewski, M., Rappl, G., Hombach, A., Riet, T., Schmidt, A., Hombach, A.A., Wendtner, C.M., and Abken, H. (2011). CD28 costimulation impairs the efficacy of a redirected t-cell antitumor attack in the presence of regulatory t cells which can be overcome by preventing Lck activation. *Mol Ther* 19, 760–767.

98. Milone, M.C., Fish, J.D., Carpenito, C., Carroll, R.G., Binder, G.K., Teachey, D., Samanta, M., Lakhal, M., Gloss, B., Danet-Desnoyers, G., Campana, D.,

Riley, J.L., Grupp, S.A., and June, C.H. (2009). Chimeric receptors containing CD137 signal transduction domains mediate enhanced survival of T cells and increased antileukemic efficacy *in vivo*. *Mol Ther* 17, 1453–1464.

99. Song, D.G., Ye, Q., Carpenito, C., Poussin, M., Wang, L.P., Ji, C., Figini, M., June, C.H., Coukos, G., and Powell, Jr., D.J. (2011). *in vivo* persistence, tumor localization, and antitumor activity of CAR-engineered T cells is enhanced by costimulatory signaling through CD137 (4-1BB). *Cancer Res* 71, 4617–4627.

100. Hombach, A.A. and Abken, H. (2011). Costimulation by chimeric antigen receptors revisited the T cell antitumor response benefits from combined CD28-OX40 signalling. *Int J Cancer* 129, 2935–2944.

101. Sanchez-Lockhart, M., Graf, B., and Miller, J. (2008). Signals and sequences that control CD28 localization to the central region of the immunological synapse. *J Immunol* 181, 7639–7648.

102. Geiger, T.L., Nguyen, P., Leitenberg, D., and Flavell, R.A. (2001). Integrated src kinase and costimulatory activity enhances signal transduction through single-chain chimeric receptors in T lymphocytes. *Blood* 98, 2364–2371.

103. Marcu-Malina, V., Van-Dorp, S., and Kuball, J. (2009). Re-targeting T-cells against cancer by gene-transfer of tumor-reactive receptors. *Expert Opin Biol Ther* 9, 579–591.

104. Patel, S.D., Ge, Y., Moskalenko, M., and McArthur, J.G. (2000). Anti-tumor CC49-zeta CD4 T cells possess both cytolytic and helper functions. *J Immunother* 23, 661–668.

105. Junghans, R., Safa, M., and Huberman, M. (2000). Preclinical and phase I data of anti-CEA designer T cell therapy for cancer: A new immunotherapeutic modality. *Proc Am Assoc Can Res* 41, 543.

106. Emtage, P.C., Lo, A.S., Gomes, E.M., Liu. D.L., Gonzalo-Daganzo, R.M., and Junghans, R.P. (2008). Second-generation anti-carcinoembryonic antigen designer T cells resist activation-induced cell death, proliferate on tumor contact, secrete cytokines, and exhibit superior antitumor activity *in vivo*: A preclinical evaluation. *Clin Cancer Res* 14, 8112–8122.

107. Parkhurst, M.R., Yang, J.C., Langan, R.C., Dudley, M.E., Nathan, D.A., Feldman, S.A., Davis, J.L., Morgan, R.A., Merino, M.J., Sherry, R.M., Hughes, M.S., Kammula, U.S., Phan, G.Q., Lim, R.M., Wank, S.A., Restifo, N.P., Robbins, P.F., Laurencot, C.M., and Rosenberg, S.A. (2011). T cells targeting carcinoembryonic antigen can mediate regression of metastatic colorectal cancer but induce severe transient colitis. *Mol Ther* 19, 620–626.

108. Kershaw, M.H., Westwood, J.A., Parker, L.L., Wang, G., Eshhar, Z., Mavroukakis, S.A., White, D.E., Wunderlich, J.R., Canevari, S., Rogers-Freezer, L., Chen, C.C., Yang, J.C., Rosenberg, S.A., and Hwu, P. (2006). A phase I study on adoptive immunotherapy using gene-modified T cells for ovarian cancer. *Clin Cancer Res* 12, 6106–6115.

109. Lamers, C.H., Sleijfer, S., Willemsen, R.A., Debets, R., Kruit, W.H., Gratama, J.W., and Stoter, G. (2004). Adoptive immuno-gene therapy of cancer with single chain antibody [scFv(Ig)] gene modified T lymphocytes. *J Biol Regul Homeost Agents* 18, 134–140.

110. Lamers, C.H., van Elzakker, P., Langeveld, S.C., Sleijfer, S., and Gratama, J.W. (2006). Process validation and clinical evaluation of a protocol to generate gene-modified T lymphocytes for imunogene therapy for metastatic renal cell carcinoma: GMP-controlled transduction and expansion of patient's T lymphocytes using a carboxy anhydrase IX-specific scFv transgene. *Cytotherapy* 8, 542–553.

111. Lamers, C.H., Langeveld, S.C., Groot-van Ruijven, C.M., Debets, R., Sleijfer, S., and Gratama, J.W. (2007). Gene-modified T cells for adoptive immunotherapy of renal cell cancer maintain transgene-specific immune functions *in vivo*. *Cancer Immunol Immunother* 56, 1875–1883.

112. Tran, C.A., Burton, L., Russom, D., Wagner, J.R., Jensen, M.C., Forman, S.J., and DiGiusto, D.L. (2007). Manufacturing of large numbers of patient-specific T cells for adoptive immunotherapy: An approach to improving product safety, composition, and production capacity. *J Immunother* 30, 644–654.

113. Till, B.G., Jensen, M.C., Wang, J., Chen, E.Y., Wood, B.L., Greisman, H.A., Qian, X., James, S.E., Raubitschek, A., Forman, S.J., Gopal, A.K., Pagel, J.M., Lindgren, C.G., Greenberg, P.D., Riddell, S.R., and Press, O.W. (2008). Adoptive immunotherapy for indolent non-Hodgkin lymphoma and mantle cell lymphoma using genetically modified autologous CD20-specific T cells. *Blood* 112, 2261–2271.

114. Pule, M.A., Savoldo, B., Myers, G.D., Rossig, C., Russell, H.V., Dotti, G., Huls, M.H., Liu, E., Gee, A.P., Mei, Z., Yvon, E., Weiss, H.L., Liu, H., Rooney, C.M., Heslop, H.E., and Brenner, M.K. (2008). Virus-specific T cells engineered to coexpress tumor-specific receptors: Persistence and antitumor activity in individuals with neuroblastoma. *Nat Med* 14, 1264–1270.

115. Kalos, M., Levine, B.L., Porter, D.L., Katz, S., Grupp, S.A., Bagg, A., and June, C.H. (2011). T cells with chimeric antigen receptors have potent antitumor effects and can establish memory in patients with advanced leukemia. *Sci Transl Med* 3, 95ra73.

116. Porter, D.L., Levine, B.L., Kalos, M., Bagg, A., and June, C.H. (2011). Chimeric antigen receptor-modified T cells in chronic lymphoid leukemia. *N Engl J Med* 365, 725–733.

117. Brentjens, R.J., Rivière, I., Park, J.H., Davila, M.L., Wang, X., Stefanski, J., Taylor, C., Yeh, R., Bartido, S., Borquez-Ojeda, O., Olszewska, M., Bernal, Y., Pegram, H., Przybylowski, M., Hollyman, D., Usachenko, Y., Pirraglia, D., Hosey, J., Santos, E., Halton, E., Maslak, P., Scheinberg, D., Jurcic, J., Heaney, M., Heller, G., Frattini, M., and Sadelain, M. (2011). Safety and persistence of adoptively transferred autologous CD19-targeted T cells in patients

with relapsed or chemotherapy refractory B-cell leukemias. *Blood* 118, 4817–4828.

118. Morgan, R.A., Yang, J.C., Kitano, M., Dudley, M.E., Laurencot, C.M., and Rosenberg, S.A. (2010). Case report of a serious adverse event following the administration of T cells transduced with a chimeric antigen receptor recognizing ERBB2. *Mol Ther* 18, 843–851.

119. Brentjens, R., Yeh, R., Bernal, Y., Riviere, I., and Sadelain, M. (2010). Treatment of chronic lymphocytic leukemia with genetically targeted autologous T cells: Case report of an unforeseen adverse event in a phase I clinical trial. *Mol Ther* 18, 666–668.

120. Newrzela, S., Cornils, K., Li, Z., Baum, C., Brugman, M.H., Hartmann, M., Meyer, J., Hartmann, S., Hansmann, M.L., Fehse, B., and von Laer, D. (2008). Resistance of mature T cells to oncogene transformation. *Blood* 112, 2278–2286.

121. Newrzela, S., Cornils, K., Heinrich, T., Schläger, J., Yi, J.H., Lysenko, O., Kimpel, J., Fehse, B., and Von Laer, D. (2011). Retroviral insertional mutagenesis can contribute to immortalization of mature T lymphocytes. *Mol Med* 17, 1223–1232.

122. Savoldo, B., Rooney, C.M., DiStasi, A., Abken, H., Hombach, A., Foster, A.E., Zhang, L., Heslop, H.E., Brenner, M.K., and Dotti, G. (2007). Epstein Barr virus-specific cytotoxic T-lymphocytes expressing the anti-CD30zeta artificial chimeric T-cell receptor for immunotherapy of Hodgkin´s disease. *Blood* 110, 2620–2630.

123. Stewart-Jones, G., Wadle, A., Hombach, A., Shenderow, E., Held, G., Fischer, E., Kleber, S., Stenner, F., Bauer, S., McMichael, A., Knuth, A., Abken, H., Hombach, A., Cerundolo, V., Jones, E., and Renner, C. (2009). Rational development of high affinity T-cell receptor-like antibodies. *Proc Natl Acad Sci USA* 106, 5784–5788.

124. Kershaw, M.H., Wang, G., Westwood, J.A., Pachynski, R.K., Tiffany, H.L., Marincola, F.M., Wang, E., Young, H.A., Murphy, P.M., and Hwu, P. (2002). Redirecting migration of T cells to chemokine secreted from tumors by genetic modification with CXCR2. *Hum Gene Ther* 13, 1971–1980.

125. Cheadle, E.J., Riyad, K., Subar, D., Rothwell, D.G., Ashton, G., Batha, H., Sherlock, D.J., Hawkins, R.E., and Gilham, D.E. (2007). Eotaxin-2 and colorectal cancer: A potential target for immune therapy. *Clin Cancer Res* 13, 5719–5728.

126. Overwijk, W.W., Theoret, M.R., Finkelstein, S.E., Surman, D.R., de Jong, L.A., Vyth-Dreese, F.A., Dellemijn, T.A., Antony, P.A., Spiess, P.J., Palmer, D.C., Heimann, D.M., Klebanoff, C.A., Yu, Z., Hwang, L.N., Feigenbaum, L., Kruisbeek, A.M., Rosenberg, S.A., and Restifo, N.P. (2003). Tumor regression and autoimmunity after reversal of a functionally tolerant state of self-reactive CD8+ T cells. *J Exp Med* 198, 569–580.

127. Kieback, E., Charo, J., Sommermeyer, D., Blankenstein, T., and Uckert, W. (2008). A safeguard eliminates T cell receptor gene-modified autoreactive T cells after adoptive transfer. *Proc Natl Acad Sci USA* 105, 623–628.

128. Torikai, H., Reik, A., Liu, P.Q., Zhou, Y., Zhang, L., Maiti, S., Huls, H., Miller, J.C., Kebriaei, P., Rabinovitch, B., Lee, D.A., Champlin, R.E., Bonini, C., Naldini, L., Rebar, E.J., Gregory, P.D., Holmes, M.C., and Cooper, L.J. (2012). A foundation for "universal" T-cell based immunotherapy: T-cells engineered to express a CD19-specific chimeric-antigen-receptor and eliminate expression of endogenous TCR. *Blood* Apr 24.

129. Pegram, H.J., Jackson, J.T., Smyth, M.J., Kershaw, M.H., Darcy, P.K. (2008). Adoptive transfer of gene-modified primary NK cells can specifically inhibit tumor progression *in vivo*. *J Immunol* 181, 3449–3455.

130. Kruschinski, A., Moosmann, A., Poschke, I., Norell, H., Chmielewski, M., Seliger, B., Kiessling, R., Blankenstein, T., Abken, H., and Charo, J. (2008). Engineering antigen-specific primary human NK cells against HER-2 positive carcinomas. *Proc Natl Acad Sci USA* 105, 17481–17486.

131. Lamers, C.H., Sleijfer, S., Vulto, A.G., *et al.* (2006). Treatment of metastatic renal cell carcinoma with autologous T-lymphocytes genetically retargeted against carbonic anhydrase IX: first clinical experience. *J Clin Oncol* 24, e20–e22.

Chapter 5

T-Cell Engineering
and Expansion — GMP Issues

Cor Lamers and Ryan Guest

5.1 Introduction

Adoptive T-cell therapy (ACT) for cancer is based on the *ex vivo* generation or selection of tumor-reactive T lymphocytes and their activation and numerical expansion before reinfusion to the autologous tumor-bearing patient. Initial approaches to ACT for cancer made use of *ex vivo* IL-2-activated peripheral blood mononuclear cells that exhibited so-called lymphokine-activated killer (LAK) activity. LAK activity was defined as the capability to kill fresh and cultured tumor cells, but not normal cells, in an MHC-unrestricted manner. Patients with advanced melanoma and renal cell carcinoma were treated with high-dose IL-2 in combination with up to 2×10^{11} autologous lymphocytes with LAK activity. Although complete and partial tumor regression were seen in up to one-third of patients, subsequent randomized studies attributed these responses to IL-2, whilst no clear evidence for a benefit of LAK cells was demonstrated.[1,2]

An alternative method for providing patients with tumor-specific T-cell immunity is the isolation, expansion, and reinfusion of tumor-specific T cells from individual patients. This approach was first shown to be feasible and render clinical successes with virus-specific T-cell immunity, i.e., with cytomegalovirus-specific, and Epstein–Barr virus-specific T cells to establish anti-viral immunity *in vivo*.[3,4] Tumor-specific T cells were first isolated from tumor tissue, i.e., tumor-infiltrating lymphocytes (TILs) obtained from melanoma patients. Adoptive transfer of bulk TIL cultures or cloned tumor-specific T cells isolated from these TIL cultures showed that this approach indeed can transfer cell-mediated anti-melanoma activity to patients.[5–7] However, the real breakthrough in TIL treatment only came after introduction in the treatment protocol of a patient conditioning regiment prior to reinfusion of TIL.[6] Using these optimized protocols, now up to 70% clinical responses have been reported in patients suffering from metastatic malignant melanoma who receive TIL treatment following preconditioning.[8–10]

However, thus far, TIL treatment is restricted to treatment of malignant melanoma in a limited number of specialized centers. The clinical application of adoptive transfer of tumor-specific cytotoxic T lymphocytes (CTLs) for cancer treatment (other than melanoma) is severely hampered by the difficult process required to reproducibly isolate and expand the TILs, either *in vitro* or *in vivo*. An alternative strategy to create large numbers of tumor-specific T cells is CTL "retargeting" by externally loading a polyclonal population of activated CTL with a bispecific monoclonal antibody (mAb) or by the transfer of genes encoding tumor-specific receptors for permanent retargeting.[11]

The initial approach to "retargeting" CTL is by using bispecific mAb. Bispecific mAbs are directed toward a tumor-associated antigen (TAA) on the one hand and toward a T-cells activating antigen, i.e., CD3, on the other, and directs the tumor cell killing potential of any CTL toward the TAA expressing tumor cell.[12] We and others have previously tested human T lymphocytes sensitized with such bispecific mAbs in clinical studies.[13,14] In spite of objective clinical responses, the use of these bispecific mAbs is complicated by a limited accessibility of solid tumors to (bispecific) mAb, dissociation of bispecific mAb from CTL, development of human anti-mouse antibody responses, and limited recycling capacity of cytolysis by mAb-sensitized T lymphocytes. In addition, the clinical anti-tumor effects were only loco-regional, and not systemic.[15–17]

One way to overcome the limitations of the bispecific mAb approach is the permanent genetic programming (retargeting) of the T-cell specificity by introduction of antigen-specific receptor genes into the patient's T lymphocytes.[11,18-21] These receptors can either be based on classical antibodies recognizing TAA, providing an MHC non-restricted receptor format, or on T-cell receptors (TCRs) isolated from a tumor-specific CTL, and thus providing an MHC-restricted format[22,23] and is further detailed in previous chapters.

All the cancer-focused adoptive T-cell immunotherapy approaches described above are based on the large-scale *ex vivo* manipulation of autologous patient T cells. These approaches can be broken down into procedures for isolation, activation, gene modification (in case of genetic retargeting), and numerical expansion before reinfusion into the tumor-bearing patient. Clinical-scale preparation methods have been defined and validated for preparation of lymphocytes with LAK activity[24,25]; virus-specific T cells[3,26]; bispecific mAb-coated activated T cells[27]; tumor-specific T cells from TIL[28-30]; and gene-modified tumor-specific T cells.[31,32]

Most of these adoptive immunotherapy approaches rely on closed cell culture technologies meeting principles of good manufacturing practice (GMP).[31-39] Parallel to development and extension of the scope of cellular products for patient treatment, regulation came in place for manufacturing these products and patient treatment.[34,40,41]

In this chapter, we will discuss the various aspects of T-cell engineering and expansion for clinical application.

5.2 Clinical Vectors

5.2.1 Gene transfer technologies

Pivotal to the success of genetic engineering of T-cell specificity is the availability of adequate gene transfer technologies, comprising both virus and non-virus-based transfer methods and are largely discussed in Chapter 2.

In short, to date, the use of γ-retroviruses is most widespread in clinical T-cell engineering, but bears some drawbacks, as it requires the induction of cell replication to allow vector integration and there may be more safety concerns associated with γ-retroviruses than with lenti-viruses.[42] Lentivirus-derived vectors are considered more efficient for gene transfer, as they can integrate into the genome of non-dividing cells[43,44] and they are less susceptible to gene silencing by host restriction factors.[45] First protocols with engineered T cells using vesicular stomatitis virus G glycoprotein (VSV-G) pseudotyped lentivirus vectors have entered clinical testing.[42,46] Promising new developments are also reported in foamy virus-derived vectors[47,48] for engineering T cells and in the field of non-virus-based transfer methods, using transposon-based methods and electroporation.[49] Various transposon and RNA electroporation-based systems are now first in clinical trials to test the safety and feasibility of this approach to engineer T cells.[50–55]

5.2.2 Clinical vectors and vector preparations

The progress in vector development for clinical cell modification is detailed in Chapter 2, and only will be discussed briefly.

5.2.3 Retroviral vectors

To date, the most commonly used vectors in the clinical practice of engineered T cells are gamma retroviral MFG(-derived) vectors,[56–58] and contain moloney murine leukemia virus (MoMLV) or murine stem cell virus (MSCV) long terminal repeats (LTRs), and optimized splicing and start codons.[59–62] It has been recognized that transduction efficiencies differ substantially between different vectors, with a major role for the viral origin of the LTRs and splice and start codon sequences. In this respect, it is of interest to mention that the MP71 vector, which has a myeloproliferative sarcoma virus (MPSC) LTR and optimal 5′ sequences, demonstrated highly improved TCR$\alpha\beta$ transduction efficiencies.[63,64] In contrast, in order to improve on the safety profile of retroviral vectors, self-inactivating (SIN) retroviral vectors are developed, using "defective" LTRs and requiring incorporation of tissue/cell specific-promoters.[65]

For TCR gene transfer, the TCRα and TCRβ genes are preferentially introduced in the host T cell as a single construct, containing both the TCRα and TCRβ chain separated by either an internal ribosomal entry site (IRES) or the 2A peptide sequence. The 2A peptide sequence is preferred for future studies, as the IRES may result in lowered expression of the downstream gene relative to the upstream gene. Thus far, in clinical studies, both IRES and 2A sequences have proved valid to separate TCRα and β genes.[62] The TCR transgene constructs and transgene cassette may affect both the efficacy and the safety of TCR gene transfer.[66] Placing the TCRβ chain in front of the TCRα chain, especially when separated by the 2A sequence, demonstrated optimal functional TCR expression levels in most TCRs tested.[67]

Different viral envelopes might be used to modify the tropism of the vector. Viral envelopes for pseudotyping vectors include the glycoprotein G VSV, the gibbon ape leukemia (GaLV) MoMLV-10A1, amphotropic MLV (MLV-A), and lymphocytic choriomeningitis virus (LCMV).[68–72]

5.2.3.1 Packaging cells

A limited number of virus producer cells, so-called packaging cells, are available for production of clinical batches of retrovirus. The PA317 cell line[73] and GP+env AM12 cell line,[74] producing amphotropic viruses were used in early clinical studies.[75–77] At present, so-called third-generation packaging cell lines are widely used, such as the murine PG13 cell line allowing for the production of MoMuLV particles pseudotyped by GaLV SEATO-strain envelope[78] and the Phi-NX-Ampho (Phoenix-A) packaging cell line, derived from human 293T cells producing amphotropic viral particles,[79] and can be used to produce clinical batches of SIN γ-retroviral vectors based on transient transfection methods.[80] Recently, a PG13-derived cell line PG368[65] was reported for the stable production of SIN retroviral vectors. Promising for scale-up for clinical-scale production is the generation of 293-derived packaging cell lines that grow in suspension in serum-free medium and produce high-titer RD114- and GALV-pseudotyped vectors.[81]

To generate a stable packaging cell line, packaging cells are transfected by the transgene containing vector plasmid or transduced with transgene containing retroviruses, produced in a transient (e.g., 293T) virus production system.[82] This is followed by selection and limiting dilution culture for generating individual high producer clones and the generation of a master and/or working cell bank.[31,83] Generation of the stable packaging cells may also be contracted out to the virus vector production facility, having available GMP-certified "empty" packaging cells, which reduces the extent of the required safety testing (see below) after having generated the packaging cell line of interest.[84]

5.2.3.2 Retroviral vector production and storage conditions

Optimal conditions for production of the clinical-grade retrovirus containing culture supernatant (RTVsup) has to be defined for each packaging cell line. For example, the optimal medium and production conditions for PG13 (clone 1.2 SFG-CAIX-CAR) were defined to be in RPMI1640 medium supplemented with either 2% human serum albumin or 10% "clinical-grade" FCS and produced at 32°C. Production in synthetic serum-free medium (e.g., AIMV) appeared inferior.[31] Optimal conditions to produce Phoenix-A (clone 58 SFG-CAIX-CAR) RTVsup were found to be similar to those defined previously for PG13 (clone 1.2) except for the production medium. Phoenix-A (clone 58)-derived RTVsup providing the highest transduction potency when using DMEM or RPMI-1640 production medium supplemented with 10% "clinical-grade" FCS.[83]

Lamers *et al.* decided to generate the clinical batch of Phoenix-A (clone 58) RTVsup in RPMI-1640 + 10% clinical-grade FCS, as the T cells to be transduced with this RTVsup are cultured in RPMI-1640 based medium.[32] Regulatory restraints might require serum-free RTVsup production. In order to achieve appropriate virus production under serum-free conditions, packaging cell lines will have to be selected/adapted for these culture conditions.[81]

For economic clinical production, RTVsup production can be harvested up to 3 times in 72 h, from a single Phoenix-A producer cell layer; these individual RTVsups may be stored at 4°C for 24–48 h without loss of potency in order to pool these harvests.[32] It is recommended that the viral sup is only thawed once between production and final use, as thawing and refreezing the virus once can reduce the viral titer between 2- and 10-fold. During controlled storage at −80°C, the crude RTVsup is stable for many years.[85] Published data for PG13-derived RTVsup and Clinical lot Phoenix-A RTVsup have been extended to 12 and 8 years, respectively (see Figure 5.1, panels B and D).

PG13 Clone 1.2 (SFG-CAIX-CAR) Retroviral Supernatant

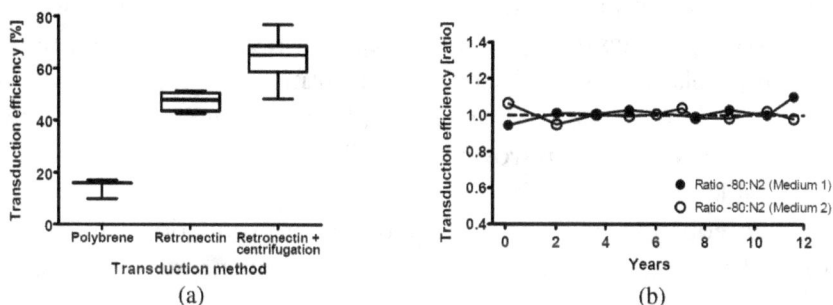

(a)

(b)

Phoenix-A Clone 58 (SFG-CAIX-CAR) Retroviral Supernatant

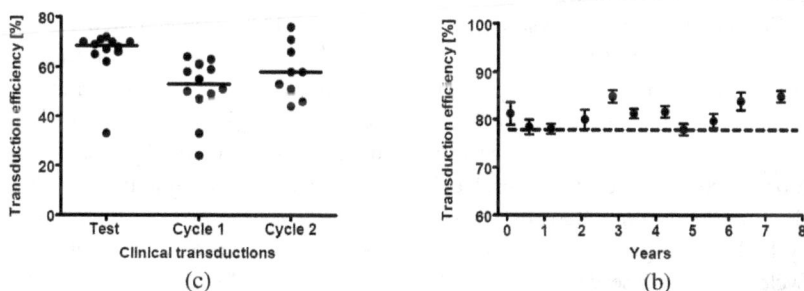

(c)

(b)

Figure 5.1. (A) Increased transduction efficiencies of retrovirus containing culture supernatants (RTVsup) as function of transduction protocol. Healthy donor PBMC were activated and transduced using PG13 clone 1.2 (SFG-CAIX-CAR) derived RTVsups produced in RPMI-1640 supplemented with either 2% HSA (medium 1) or 10% FCS (medium 2) and stored at −80°C or at −196°C in liquid nitrogen (N2). Transduction protocols used: (i) polybrene, period 1998, (ii) Retronectin®, no centrifugation, period 1999–2000; (iii) Retronectin® with centrifugation, period 2001–2010. Transduction efficiency is determined by flow cytometric assessment of the proportion [%] CAIX-CAR expressing T cells, using the anti-Id G250 mAb NuH82. Results are for protocol (i) median 14% (range: 10–17%), protocol (ii) 47% (43–51), and protocol (iii) 64% (48–77). Differences between protocols are highly significant (t-test p-values <0.001). (B) PG13 clone 1.2 (SFG-CAIX-CAR) derived RTVsups stored at −80°C are stable for up to 12 years. Transduction efficiencies determined for four different types of RTVsup (see legend to Figure 5.1A) were

5.2.3.3 *Vector batch release/acceptance criteria*

The packaging cell master/working cell bank can apply for GMP certification after demonstration of (i) intact and functional integration of the introduced transgene (i.e., intactness, copy number, and stability, as determined by full DNA sequencing, Southern blot analysis, and flow cytometry (the latter in short- and long-term cultures, i.e., >100 passages)), (ii) sterility (negative tests for bacteria, fungi, and mycoplasma), (iii) absence of adventitious viruses (assessed in extended *in vitro* and *in vivo* testing), human viruses (by PCR), and replication competent retroviruses (RCR) (extended PG4 S+L-assay) (see Table 5.1).

Subsequently, a clinical-scale pilot vector production will be performed to check upscaling of culture conditions and verify logistic

---

Figure 5.1. (*Continued*) expressed relative to N2 storage variants −80°C vs N2 for HSA supplemented RTVsup (closed circles) and FCS supplemented RTVsup (open circles). Dashed line, ratio = 1.0. (C) Efficiencies of clinical transductions. PBMC obtained from 12 RCC patients treated in the Rotterdam RCC trial were transduced with the Clinical lot of Phoenix-A clone 58 (SFG-CAIX-CAR) RTVsup. Prior to patient inclusion, a test transduction was performed (method: spinoculation on Retronectin®-coated wells; labeled "Test"). Clinical transductions (method: static transduction in Retronectin®-coated bags) for the first treatment cycle ("Cycle 1") was performed using fresh PBMC, for the second treatment cycle ("Cycle 2") cryopreserved and thawed PBMC were used. PBMC were purified by ficoll separation from blood (test transduction) or apheresis product (clinical transductions). Individual and median (line) transduction efficiencies are shown; test transduction: median 69% (range: 33–72%, $n = 12$), cycle 1: 53% (24–64, $n = 12$) and cycle 2: 58% (44–76, $n = 9$). Transduction efficiencies for the test transduction are significantly higher than for cycle 1 (paired t-test $p = 0.004$) and cycle 2 (paired t-test $p = 0.01$). Cycle 1 vs cycle 2: not significant. (D) Clinical lot of Phoenix-A clone 58 (SFG-CAIX-CAR) RTVsup displays stable transduction efficiencies upon long-term (to date 8 years) controlled storage at −80°C. About 11 transduction experiments were performed in a period up to 8 years (2002–2010) since RTVsup production, using cryopreserved PBMC of the same five healthy donors. Method spinoculation on Retronectin®-coated wells. The transduction efficiencies are presented as mean percent (±SD) of CAIX-CAR expressing T cells. Dashed line, represents the mean % of transduced T cells at 1 month since RTVsup production, i.e., 83% CAIX-CAR+ T cells.

Table 5.1. Quality control and safety tests performed on the clinical lot of Phoenix-A clone 58 (SFG-CAIX-CAR) retroviral culture supernatant and clinical lots of transduced patient T lymphocytes.

Test	Working cells	Cell bank supernatant	Clinical batch[a] supernatant	EPC	Process validation[b] TLD	Clinical protocol TLP
Identity						
– Southern blot: integration/integrity	X	—	—	—	—	—
– Sequencing: integrity	X	—	—	—	—	—
– FACS analysis[c]						
– Relevant gene expression	X	—	—	—	X	X
– Potency supernatant (transduction efficiency)	—	X	X	—	—	—
Sterility						
– Bacterial, yeast, fungal contamination	X	—	X	—	X	X
– mycoplasm	X	—	X	—	—	—
– Detection of virus						
– *In vitro*: 3 detector cell lines	X	—	X	—	—	—
– *In vivo*: suckling mice/	X	—	—	—	—	—
– embryonated eggs						
– Mouse antibody production	X	—	—	—	—	—
– XC plaque assay	—	X	—	—	—	—
– Human viruses (PCR)[d]	X	—	—	—	—	—

(*Continued*)

Table 5.1. (*Continued*)

Test	Working cells	Cell bank supernatant	Clinical batch[a] supernatant	EPC	Process validation[b] TLD	Clinical protocol TLP
Replication competent virus[c]						
– Co-cultivation/Amplification on Mus dunni followed by PG4 S+L- focus assay	X	X	X	X	X	—[f]
Endotoxin						
– LaL test	—	—	X	—	—	—
Viability						
– Trypan blue test	—	—	—	—	X	X

Abbreviations: EPC: end of production cells; TLD: transduced donor T lymphocytes; TLP: transduced patient T lymphocytes. X = test performed; — = not tested.

[a] Clinical Batch: Clinical lot of Phoenix-A clone 58 (SFG CAIX-CAR) retroviral Supernatant.

[b] Process validation: Clinical scale transduction and expansion according to the clinical protocol using lymphocytes of healthy donors.

[c] FACS analysis is performed directly on cells, i.e, packaging cells and transduced T lymphocytes or indirectly for supernatant, i.e., after supernatant transduction of primary human T lymphocytes.

[d] Tested human viruses: HIV I/II, HTLV I/II, HBC, HCV, HCMV, HHV6, and EBV.

[e] Cell numbers and volumes tested in RCR test: Working cell bank: 1% of cells, 5% of culture supernatant; Clinical batch: EPC 1×10^8 cells, supernatant 1,250 ml; TLD in process validation: 1×10^8 cells per donor.

[f] TLP cells are stored for retrospective RCR testing if clinically indicated.

aspects, followed by the GMP production of the clinical vector batch. The resulting clinical lot of retroviral supernatant has to meet the quality (identity and potency) and safety criteria (sterility, absence of adventitious viruses, RCR, and endotoxin).

Identity and potency of the clinical vector batch can be assessed on a permissive cell line (for titer assessment), or in a small-scale transduction experiments using PBMC derived from healthy donors. Subsequently, a process validation may be performed, i.e., clinical-scale transductions, in order to demonstrate (i) the validity of the clinical-scale transduction protocol in conjunction with the clinical vector batch, and (ii) that genetically modified T lymphocytes generated in this procedure are free of replication competent virus (RCR/RCL).

5.2.3.4 *Potency testing*

Assessment of the viral titer is done on permissive cell lines and depends on the envelope pseudotyping of the chosen vector, e.g., 293 cell line for GaLV-enveloped vectors and Mus dunni for MLV-enveloped virus.[86–88] Alternatively, the potency of the virus batch can be determined by assessment of the functional transduction efficiency on the final target cell, e.g., primary human T lymphocytes, without assessment of the exact titer.[32] The last method elicits very reproducible and stable results,[85] and Figure 5.1 (panel D).

5.2.3.5 *Replication competent retrovirus*

A major issue in safety testing is the proof that the vector batch is free of RCR, which might have been generated by putative recombination events. Current vectors are designed and constructed in order to reduce chances of RCR/RCL events generated by recombination events. Despite the fact that several groups have developed qPCR assays for detection of putatively generated RCR,[89–92] sensitive biological assays are prescribed to show that the volume of a clinical virus lot, to be used for the treatment of one patient, contains less than one RCR.[93] These bio-assays are based on a two-step procedure: (i) amplification of the putative retrovirus on a permissive cell line, e.g., 293 cell line for GaLV-enveloped RCR and Mus dunni for MLV-enveloped RCR followed by (ii) testing amplification cells and/or culture supernatants for RCR in the of the PG4 S+L-assay.[89,94]

5.2.4 Lentiviral vectors

Lentiviral vectors are based on HIV-1 and consist of a replication-incompetent, non-pathogenic version of the virus. The development of SIN lentiviral vectors by deleting/replacing LTR elements improved the safety of the HIV-1 derived vectors without reducing gene delivery efficiency in both dividing and non-dividing cells and enabled development of clinical applications.[95-97]

5.2.4.1 Packaging cells

There are several versions of the HIV-1-based lentiviral packaging systems; they are referred to as first, second, and third generation. The differences between these versions reside in the number of plasmids used for packaging, allowing the cis- and trans-elements to be split on different plasmids and thereby improving the biosafety profile of the vector preparations. The first and second generations consist of three plasmids: (i) a vector plasmid, (ii) a packaging plasmid and (iii) an envelope plasmid. The number of accessory genes present on the packaging plasmid varies depending on the system. The third-generation system includes four expression cassettes: (i) a vector plasmid, (ii) a gag/pol plasmid, (iii) a rev plasmid, and (iv) an env plasmid. Packaging cells line that is commonly used to produce lentiviral vectors are 293T/17 cells (ATCC#CRL-11268), a highly transfectable derivative of the 293 human embryonic kidney (HEK) cell line, into which the temperature-sensitive gene for simian virus 40 (SV40) large T antigen has been inserted.[98]

5.2.4.2 Lentiviral vector production and purification

A prerequisite for clinical application of lentiviral vectors is the development of large-scale vector production methods.[99,100] The development of a scalable process for high-yield lentiviral vector production by transient transfection of HEK293 suspension cultures enable lentiviral vector production at industrial scale in bioreactors. This allows the production of lentiviral vectors in sufficiently large quantities for phase I clinical trials.[101]

Upscaling of the transfection production system and the downstream processing is a technical challenge. Following transfection of the packaging cells with the vector plasmids, the virus is harvested following 24–36 h of incubation by collection of the culture medium. Vector-containing medium is clarified by filtration, and subsequently concentrated by ultrafiltration, resulting in the concentrated harvest. Residual cellular and plasmid DNA is removed enzymatically, followed by gel-filtration. Finally, following sterile filtration, the vector is aliquoted into plastic bags/ampoules as the final fill.

5.2.4.3 Vector batch release

The QC release testing include, in addition to assays listed for retro viral vector products (see Table 5.1), assays to eliminate contamination with vector elements, such as, envelope (VSV-G)RNA/DNA, Nef DNA E1A/E1B DNA, p24 protein, and processing-related contaminants. Assays for replication competent lentivirus (RCL) are, as depicted for retrovirus, based on a two-step procedure: (i) putative RCL are amplified by culture on a permissive cell line, e.g., SUP-T1 (ATCC#CRL-1942) lymphoblast cells,[98] C8166-45 cells,[92,102] or onto HEK 293T cells[103] and (ii) subsequently the cells used for amplification and the amplified culture supernatant are tested for cytopathic effects on indicator cells, e.g., C8166-45 or MT-4 cells, and/or confirmation by p24 Elisa to detect HIV-1 capsid and PCR for psi-gag recombination.[102]

5.2.4.4 Clinical application

The first clinical trial using a lentiviral vector was conducted in subjects with chronic HIV infection, who received autologous CD4$^+$ T cells gene-modified with an anti-sense gene against the HIV envelope.[46]

The generation of clinical lots of lentivirus using a laborious and costly transfection production system followed by a complex purification process thus far has limited the initiation of phase I/II clinical trials. However, with advances in vector design, large-scale production and purification technologies, lentiviral vectors have become a safer and more effective gene delivery system. Since the first clinical trial was approved in 2002,[104] several trials have been reported using lentivirus-transduced cells to treat patients with malignancies and infectious and genetic diseases.[46,105–108]

5.3 *Ex Vivo* Generation of Gene-Modified T Cells

The procedure for the *ex vivo* generation of gene-modified T cells include following steps: (i) the isolation of the starting cell population, in general PBMC or purified T lymphocyte (subsets), followed by (ii) activation (required for γ-retroviruses only), (iii) gene modification, and (iv) numerical expansion of the T lymphocytes before reinfusion to the autologous tumor-bearing patient.

In small-scale laboratory, testing optimal conditions have been defined for the respective steps in the procedure, in order to generate optimal functioning gene-modified effector cells. Initially, the field aimed at the generation of gene-modified T cells with optimal *in vitro* effector function, nowadays it is recognized that optimal and long-lasting *in vivo* effector function might be best realized by using less differentiated T cells.[109] These changing insights require adaptations in the current protocols, which are still under investigation.

5.3.1 *Critical reagents*

In the translation of laboratory procedures to clinically applicable protocols, several critical reagents and disposable(s) have been identified.

5.3.1.1 *Culture media*

The T-cell culture media that are currently used for clinical *ex vivo* T-cell culturing and manipulations are either standard basic media, such as RPMI 1640, or synthetic serum-free formulations such as Aim-V (Invitrogen, Grand Island, NY), X-Vivo-15 (Cambrex, East Rutherfort, NJ), and CellGro SCGM (CellGenix, Freiburg, Germany). All of these media are buffered by HEPES in combination with bicarbonate in order to maintain a strict physiological pH of 7.2 to 7.6.

The most essential amino acid required for eukaryotic cells is L-glutamine, which is the most abundant amino acid in human plasma.[110] It is required for protein production and is an alternative energy source to glucose in cell metabolism. Due to its poor stability in culture media, it has historically been used as an additive to basic media. Nowadays, most media are supplemented with Glutamax, a stable synthetic homolog of L-glutamine, which improves the shelf life of base media without the need to add L-glutamine.

The composition of synthetic serum-free medium formulations is proprietary. A side by side comparison of these media showed slight differences in cell expansion rates,[27,111] lymphocyte subset distribution,[112] and functional properties. Yet, for all serum-free medium formulations, T-cell expansion properties improved when adding human serum.[111] For economic reasons, the serum-free formulations might be blended with basal media, e.g., in ratio of 1:4 (20% AIMV+80% RPMI 1640,[27,32] or 1:1 (50% AIMV/50% RPMI 1640; M.E. Dudley, personal communication), however, blending requires the addition of 2–5% of serum or plasma. Thus, a careful evaluation of different culture media can greatly benefit the quality and quantity of the cellular product generated for immunotherapy.

5.3.1.2 *Medium additive: human serum*

While serum-free media have been investigated for some decades, there is still a requirement for serum to enable a suitable expansion of the T cells during the *ex vivo* culture process.[111] The reported literature demonstrates that fetal bovine serum (FBS) is still the gold standard for expanding T cells; however, the risk of both TSE and anti-FBS immune responses have led to the use of human serum. Preferably, human AB serum is used to reduce the likelihood of antibodies to A or B antigens, and the sera have to meet the standard safety and quality control measures for heterologous blood product donation. Alternatively, medium might be supplemented with human plasma, either from autologous source, gathered during leukapheresis, or from fresh frozen plasma packs as used in clinical practice.[27,32]

5.3.1.3 T-cell activation agents: monoclonal antibodies anti-CD3 (OKT3) and anti-CD28

In general, T cells are activated by using the anti-CD3 mAb OKT3, binding the TCRε chain, alone or in combination with the mAb binding the co-stimulatory molecule CD28 to induce T-cell proliferation.

Proliferation of the T cells is important for two reasons: (i) to allow integration of the retro-viral vector (specifically γ-retrovirus) containing the therapeutic gene(s) into the genomic DNA[113] and (ii) generate sufficient numbers of cells for therapy. It is important to note that soluble anti-CD3 mAb (or other co-stimulatory antibodies) will only stimulate T cells through the TCR proteins following cross-linking by the Fc-part of the mAb by Fc-receptor expressing cells, such as monocytes. Alternatively, immobilized anti-CD3 mAb along with co-stimulatory antibodies (and or cytokines) are a suitable alternative; the antibodies can be immobilized on either coatable surfaces such as tissue culture plastics or on beads.[31,114–116] It has to be noted that the activation agent/method might affect the final T-cell product composition regarding T-cell maturation status, base line activity, and hence could impact on the engraftment and possible clinical function of the final product.

Of note, the availability of GMP-grade murine mAb OKT-3 has been discontinued by its original supplier due to the toxicity seen during its prescribed *in vivo* clinical use. However, it has been replaced with a humanized version of the mAb and while this mAb has not yet been used in the T-cell immunotherapy clinical setting, there is no reason to assume that it will not be equally as efficient for *ex vivo* activation of T cells. However, a GMP grade of the original murine mAb OKT-3 will again become available through a different supplier for *ex vivo* applications.

5.3.1.4 Cytokine support for T-cell cultures: IL-2

Pivotal for T-cell growth and support *in vitro* is the cytokine interleukin 2 (Proleukin, hu-rIL-2, Chiron Corporation). IL-2 provides both a proliferative signal and anti-apoptotic signal to cytotoxic T cells and is currently added to cultures to sustain the cells throughout the *ex vivo* manipulation phase.[117] Clinical-grade IL-2 is available and has a long history of use in *ex vivo* T-cell expansion and clinical trials for *in vivo* application.[118]

In current protocols for the generation of gene-modified T cells, IL-2 is added during the activation and transduction process at final concentration of 100 IU/ml (however, IL-2 may be left out during the activation phase), which is increased to 360 IU/ml throughout the expansion phase. The IL-2 concentration might critically impact the T-cell maturation state of the final product.[119] To date, the effects are evaluated of other members of γ-chain family of cytokines, like IL-7, IL-15, and IL-21 on T-cell properties in the context of ACT. These cytokines show an interplay in growth and differentiation of antigen-specific T cells *in vitro*.[120,121] GMP-grade preparations of these cytokines are available or will become available shortly for clinical evaluation.

5.3.1.5 Retronectin

Retronectin®, the human fibronectin fragment CH296 (Takara, Otsi, Japan), has been demonstrated to enhance retroviral transduction of immune cells.[31,122] Retronectin® physically brings together retrovirus and immune cells. The agent binds retrovirus through the heparin-binding domain and immune cells via the CS-1 and RGDS domains that bind to T-cell activation markers VLA4 and VLA5, respectively.[123] However, Retronectin® needs to be coated on a solid phase, like polystyrene plates, dishes or flakes, for effective co-localization of retrovirus and immune cells.

5.3.2 *Starting cell population*

5.3.2.1 Autologous leukapheresis

Lymphocytes utilized for clinical investigation and therapy are most commonly harvested from the peripheral blood by leukapheresis or, in some cases, from excised tumor tissue. PBMC from leukapheresis are, however, a diverse assortment of lymphocytes, including T cells, B cells, and natural killer (NK) cells, as well as a substantial number of monocytes, to lesser extent dendritic cells (DCs), and are contaminated with neutrophils, erythrocytes, and platelets. The use of unfractionated leukapheresis product as starting material for T-cell activation and gene modification poses a high risk of cell coagulation, resulting in cell loss (up to 90%, especially from cryopreserved aphereses) either in the activation bag or during initial purification procedures. The presence of erythrocytes and platelets also has an impact on the level of transduction as they are capable of binding to Retronectin® and hence reduce available binding sites for activated T cells and/or viral particles. The use of cryopreserved aphaeresis product might be considered as starting material of choice as cryopreservation and thawing breaks down erythrocytes and neutrophils and these are removed along with platelets during the washing process prior to setting up the activation culture.

PBMC can be isolated from the leukapheresis product by a ficoll procedure; open procedures should be avoided to reduce risk of contamination. Several closed and automated cell processing devices are available, for isolation of PBMC from aphaeresis products (e.g., COBE 2991 and Elutra™ cell processing devices cell (Gambro BCT, Lakewood, CO, USA). Further isolation of lymphocytes or lymphocyte subsets are possible using the following strategies: (i) isolation of lymphocyte subpopulations using paramagnetic clinical devices (CliniMACS®, Miltenyi, Bergisch Gladbach, Germany: CE marked; Isolex 300i, Baxter: FDA approved; Dynal ClinExVivo MPC magnet (Invitrogen, Carlstad, CA); (ii) alternatively: clinical flow sorting high speed FACS within GMP facility (Gigasort TM a non-aerosol-generating microfluid sorting device; Cytonome, Boston, MA, USA[6,32–34,38,55,124,125]; (iii) elutriation: modified protocols are available for the Elutra® for the successful and efficient high-yield enrichment of lymphocytes and concomitant elimination of monocytes from leukapheresis using counter flow centrifugal elutriation.[126] The method of choice depends on the clinical protocol defining the requirements of the starting cell population.

5.3.2.2 Cryopreservation of leukapheresis product

Cryopreservation is carried out as per a standard stem cell cryopreservation for patients with hematological malignancies undergoing stem cell collections. Briefly, the initial collection is diluted to $\leq 200 \times 10^9$ nucleated cells/l using either autologous plasma (isolated during the apheresis process) or 4.5% HSA solution (Bio Products Laboratory, UK). The same autologous plasma or 4.5% HSA solution is also used to prepare the 20% DMSO freezing media which is precooled on ice prior to a 1:1 addition of the aphaeresis material. The mixture is aliquoted into suitable cryopreservation bags and frozen using a controlled rate freezer. The product can then be stored in the vapor phase of liquid nitrogen until required.

5.3.2.3 Thawing and washing the leukapheresis collection

The cryopreservation bag is warmed on a 37°C hotplate until a small quantity of ice remains. Approximately, 5–20 fold excess of cold (2–8°C) complete T-cell growth media is added to reduce the likelihood of cell damage and subsequently, the mixture is washed with cold (2–8°C) complete T-cell growth media using an Elutra™ cell processing or by centrifugation using blood bags, to provide the final cell preparation used for activation.

5.3.3 T-cell activation and expansion

5.3.3.1 Activation

Clinical T-cell activation protocols for the generation and expansion of gene-modified T cells commonly used sCD3 mAb OKT3 at a final concentration of 10–100 ng/ml to activate PBMC. Protocols might differ in whether or not IL-2 is added to this initial activation stage of the culture, from no exogenous IL-2,[31,32] to low dose 25–100 IU/ml,[34,55,127] or high dose 1,000 IU/ml.[128] The higher the IL-2 concentration at the initial stage of the culture, the more non-antigen-specific activated kill (AK) activity is induced in the T-cell population.[31]

When T cell cultures are started with fractionated PBMC, either by depleting monocytes, or by enriching for T cells or T-cell subsets, the activating antibody needs to be supplied on an artificial cross-linking matrix, which is provided by coating the antibodies on plates or to beads. In clinical practice, enriched T-cell subsets are activated by mAbs coated on beads, and combining anti-CD3 mAb with a co-stimulating mAb anti-CD28 mAb. Using paramagnetic CD3/CD28-coated beads allows for simultaneous T-cell isolation and activation, and facilitates removal of the beads at the end of the culture process.[34,128]

5.3.3.2 *Interleukin-2*

From day 1–2 after anti-CD3/CD28 mAb activation onwards, IL-2 is needed to sustain the T-cell cultures, however, in contrast to TIL cultures that are routinely expanded at the high dose of 6,000 IU/ml IL-2,[6] protocols for the generation of gene-modified T cells vary in the applied IL-2 concentrations between low, i.e., 25–50 IU/ml,[55,127] and intermediate, typically ranging between 300 and 600 IU IL-2/ml.[31,32] Re-evaluation of IL-2 concentration in TIL cultures (standard at 6,000 IU IL-2/ml) showed that modifying the IL-2 concentrations during culture, i.e., 10–120 IU/ml at the initiation followed by 600–6,000 IU/ml from the second week of culture onwards improved T-cell function. Reducing IL-2 concentrations in the expansion phase from 6,000 to 600 IU IL-2/ml did not affect TIL properties.[129]

5.3.3.3 *Rapid expansion*

Following T-cell activation through CD3 mAb TCR interaction, the IL-2 supported T-cell expansion cultures slow down their vigorous expansion after about 10–14 days. In a part of the trials using T cells gene modified by transduction and followed by a short-term (2–3 week) culture, this feature is not limiting the clinical application.[32] However, in antigen-specific T-cell therapy, based on selection of naturally occurring virus/tumor-antigen-specific T cells (including TIL), or adoptive gene-modified T-cell therapy based on transfection protocols, followed by (lengthy) selection procedures, this feature hampers the ability to generate sufficient cell numbers for therapy.

Riddell *et al.*[3] have developed a rapid expansion method for virus-specific T cells, based on repetitive stimulation of the T cells with CD3 mAb OKT3, IL-2, and feeder cells. Feeder cells are composed of irradiated allogeneic PBMC and allogeneic EBV-transformed B-lymphoid cell lines (LCLs). In this system, the CD3 mAb provides the TCR-mediated signal (signal 1), whereas the interaction with the feeder cells provides the co-stimulatory signal (signal 2).

The rapid expansion method has been applied for phase I/II adoptive immunotherapy studies in lymphoma, glioma, neuroblastoma, and viral diseases.[33,55,130–132] In addition, rapid expansion methods have also been allied for TIL and TCR-gene-modified T cells following transduction, as a shorter culture times are preferred, not only to reduce on labor and cost, but also to improve quality and *in vivo* function of the therapeutic cells.[6,8,62]

5.3.3.4 Artificial antigen-presenting cells

In a natural immune response, DCs as the "professional" antigen presenting cells (APCs) provide both the primary and co-stimulatory signal to the T cell in order to generate a full T-cell effector function and long-lasting immune response. These signals can also be provided through artificial APC (aAPC). The first-generation aAPCs were magnetic beads coated with clinical-grade human anti-CD3 and anti-CD28 mAbs. These aAPC induced efficient growth of human polyclonal naive and memory $CD4^+$ T cells.[116] The expanded T cells retained a highly diverse TCR repertoire and could be induced to secrete cytokines characteristic of Th1 or Th2 cells, depending on the culture conditions.[133] In addition, no exogenous growth factors or accessory cells are needed to enable T-cell stimulation and expansion.

Recently, cellular aAPC lines, derived from the chronic myelogenous leukemia line K562 have been described.[125] K562 cells do not express MHC or co-stimulatory ligands. The K562 aAPC have been transduced to express the co-stimulatory molecule 4-1BB and the high affinity Fc receptor CD64. The latter allows the flexibility of loading antibodies directed against T-cell surface receptors. CD3 and CD28 mAbs added to the K562 aAPCs cells are bound by the Fc receptor to yield a cell that expresses anti-CD3 mAb, anti-CD28 mAb, and 4-1BB.[125] These cell-based aAPCs have proved to be more efficient in activating and expanding T cells, especially CD8+ subsets and antigen-specific T cells, than the magnetic bead-based aAPC. In addition, the K562 aAPCs are capable of stimulating CD4 cells efficiently.[125]

This cellular aAPC method bears great promise as it eliminates the need for the allogeneic feeder cells (PBMC and B-LCL), but its use is costly and impractical in larger trials where the production of large numbers of patient-specific product is required. The cellular aAPC can be manufactured as a master cell bank according to GMP requirements and be used for phase I/II trials.

5.3.4 *Translation of T-cell cultures to clinical scale*

Manufacturing of large numbers of gene-modified cells for adoptive T-cell therapy is a technological and logistical challenge. In general, all lymphocyte-processing steps are performed in clean rooms (GMP laboratory) using closed cell processing to improve the safety profile of the process and comply with regulations regarding cell and tissue processing. To date, not all centers performing clinical trials in ACT have adopted these closed cell processing technologies for the activation and transduction of T cells, and generally start using standard tissue culture flasks and plates followed by expansion in cell culture bag systems. The choice for the methodology is also dictated by the available cell number at the start, and the final cell number required for the clinical protocol. However, by the successful translation of the

T-cell activation and transduction protocols from open (plate) to closed (bag) cell processing can be applied during all phases of the manufacturing process. Regulations do not prohibit open cell preparation procedures; however, these preparations have to be performed in laminar flow cabinet (class A) in class B/C clean room.

In order to justify the methods/environment used, a suitable assessment should be carried out to define how the following risk factors will be suitably controlled: Contamination of other therapeutic products with patient-derived infectious agents; Cross-contamination of one patent's therapeutic products with another's; infection of patient with RCR derived from production of the gene vector or recombination with retroviral gene sequences in patient cells; tumorigenic events in the patient due to use of recombinant retrovirus; presence of causative agents of transmissible spongiform encephalopathies (TSE); adverse immunological and toxicological effects from residual raw materials and reagents from processing; infection of patient cells with microorganisms derived from materials, staff, and the environment; and cross-contamination of other therapeutic products with retroviral gene vector.

5.3.5 Closed-system cell culture

At present, there are two closed cell culture bag platforms, i.e., the gas permeable culture bags for static culture and the wave bioreactors for continuous perfusion culture. The bags comprise injection sites for safe (needleless) access, sterile connection device (TSCD; Terumo Medical, Elkton, MD), compatible integral tubing sets with spike and luer Lock adapters, and some brands include sampling pouches. Closed bag culture technology relies on sterile fusion (TSCD) of tubing for asceptic, sampling, adding media and supplements, subculture, wash procedures. Actions involving injection sites have to be performed in a laminar flow cabinet (again a grade A environment with a grade B\C background). Gas-permeable bags are available from several suppliers, e.g., LifeCell PL732 bags (Baxter); Origon Permalife PL240 bags (Origen Biomedical, Austin, TX); cell differentiation bags (Miltenyi), and VueLife AFc bags (CellGenix, Freiburg, Germany).[31,32,36,38,134]

The wave bioreactor (Wave 2-10E; Wave Biotech LLC, Somerset, NJ), and wave EHT bioreactor (GE Healthcare, Somerset, NJ) are perfusion bioreactor systems that employ a disposable sterile culture bag that can be readily accessed, has medium perfusion capability, and can be configured to meet the clinical process requirements. The disposable culture bag is attached to rocking platform and gassed with 5% CO_2. Typically, the rocking speed of the bioreactor is set at 10 to 12 rocks per minute at an angle of 4° and the temperature at 37°C. The wave bioreactor system using medium perfusion is promising tool for manufacturing up to 6×10^{10} human T cells at cell densities reaching 20×10^6 cells/ml in a single bag. The expanded T cells remain biologically functional and can be re-activated to produce high amounts of cytokines. Wave cultured gene-modified cells have already been applied in clinical trials.[34,38]

5.3.6 Gene modification

5.3.6.1 Retroviral transduction

Transduction of T cells was first done by co-culture of activated T cells with viral packaging cells, which lead to good transduction efficiencies.[135] However, the risk of contamination of the final cell product with these packaging cells has precluded the application of this approach in clinical protocols. Retroviral transduction with virus containing culture supernatant RTVsups was effective in presence of polycations such as polybrene or protamine sulfate, presumably by increasing virus adsorption and/or membrane fusion through a charge-mediated mechanism.[136] Furthermore, an additional centrifugation step at $1,000 \times g$ further increased transduction efficiencies in the presence of these polycations.[71]

Yet, there was still room for improvement and Hanenberg and colleagues showed that co-localization of retroviral particles and cells on the human fibronectin fragment (CH296 or Retronectin) substantially increased transduction efficiency of hematopoietic cells, including primary T cells of human and murine origin[122,123,137] However, a prerequisite for the effective co-localization of retrovirus and immune cells is that Retronectin is coated on a solid surface, such as polystyrene plates, dishes, and flasks. As was previously shown for polycations, transduction of T cells on Retronectin-coated plates can be substantially enhanced by centrifugation, and this "spin-transduction" on Retronectin-coated plates has become the standard laboratory T-cell transduction protocol[138,139] (see also Figure 5.1A).

5.3.6.2 Translation to clinical-scale transduction

We optimized a protocol for gene transduction and expansion of human T lymphocytes for ACT of renal cell cancer patients with autologous CAIX-CAR gene-modified T cells (see Table 5.2).[31] GMP criteria and cost effectiveness favor to minimization of the volume of the RTVsup and maximization of T-cell density per transduction. Optimal conditions are defined as 0.5×10^6 lymphocytes per 0.3 ml of RTVsup per square centimeter of Retronectin-coated ($3\,\mu g/cm^2$) surface. The optimal window for retroviral transduction is between 1 and 3 days following anti-CD3 mAb T-cell activation. A single Retronectin-assisted transduction at day 2 after T-cell activation yields high (typical 30–60%) transduction efficiencies. Further improvement of transduction efficiency is obtained by tranducing twice (day 2 and day 3; a second transduction improves transduction efficiency by about 10%); precoating of the Retronectin-coated plate with RTVsup; including centrifugation during precoating of the plate with RTVsup and following addition of the T cells (centrifugation improves transduction efficiency with another 10%).

The defined protocol, excluding the centrifugation step, was succesfully translated to clinical scale. The bag transduction method initially was set up using the Nexell Lifecell® X-fold™ cell culture container which has a coatable polystyrene inner surface (see Figure 5.1C).[31,32] As this bag is no longer available from February 2007, alternatives have been evaluated, amongst others: Cell Differentiation bag (Miltenyi Biotec, Bergisch Gladbach, Germany), CultiLife Spin bags (Takara Bio Inc., Otsu, Japan), and VueLife™ bags AFC, Gaithersburg, MD, USA). Side by side comparison showed

Table 5.2. Details of GMP transduction protocol (Rotterdam, The Netharlands/Manchester, UK).

	Cell culture bag	Transduction bag
Day 0 Activation of PBMC	Activate PBMC at $1*106$ cells/mL with 10 ng/mL anti-CD3 mAb OKT-3 in 400–600 mL complete MixMed in 1,000 mL culture bag.	
Day 1 Retronectin coating		DC generation bag (volumes and cell numbers for a 90 cm^2 bag). Add 3 mg Retronectin/0.25 mL saline/cm^2. Incubate overnight at 4°C.
Day 2 Retroviral transduction (1)	Harvest activated PBMC and wash three times in MixMed medium without supplements, resuspend in minimal volume, count.	Rerrove retronectin. Add 30 mL blocking buffer, i.e., saline with 2% human serum albumin (NaCI/2%HSA) and incubate 1 h at 37°C. Remove blocking buffer and rinse with NaCL/2%HSA. Add 31 mL RTVsup + 100 IU/mL IL-2. Incubate 30 min at 37°C and 5% CO_2. Add $0.5*108$ PBMC to the bag (volume 1–5 mL: use syringe) ≈ 0.33 mL RTVsup + $0.55*106$ PBMC per cm^2. Incubate 6 h at 37°C and 5% CO_2. Add 37.5 mL RPMI-1640 + 8% (autologous) Human serum. Incubate overnight at 37°C and 5% CO_2.

(Continued)

Table 5.2. (*Continued*)

Day 3 Retroviral transduction (2)	Remove supernatant, leave ± 10 mL. Centrifuge to recover the cells from the supernatant. Resuspend cell pellet in 31 mL RTVsup + IL2 Add RTVsup + cells to the bag. Incubate 6 h at 37°C and 5% CO_2. Add 37.5 mL RPMI-1640 + 8% (autologous) Human serum. Harvest cells from the bag; determine cell recovery.
Day 4 Cell culture	Culture cells in complete MixMed-medium + 600 IU IL-2 (0.25*106 cells/mL) in 1,000 mL culture bag.
Day 7, 9,11 Cell culture	Determine cell growth and transduction efficiency (flow cytometry). Culture cells in complete MixMed-medium + 600 IU IL-2 (0.25*106 cells/mL) in 1,000 mL or 3,000 mL culture bags.
Day 14	Determine cell growth and transduction efficiency (flow cytometry). Determine transgene-specific functions and activated kill (AK) activity.

MixMed medium: 20% AIMV + 80% RPMI-1640 + supplemented with 2 mM L-glutamine and 50 mg/mL gentamycin.
Complete MixMed: MixMed medium supplemented with 2 IU/mL heparin and 2% human plasma.
Saline: 0.9% NaCl solution.
RTVsup: Retrovirus containing culture supernatants.
Transduction bag: DC generation bag (Miltenyi) available at 90, 180, and 270 cm² (Till February 2007: X-fold bag. Baxter). Cell culture bag: Life cell PL732 (Baxter; no longer available from May 2010; expiry date last lot: 2015). Bag manipulations are performed using sterile tube welding and heat seals; actions using syringes are performed in Laminar flow cabinet.

that Cell Differentiation bags as well as CultiLife Spin bags, when used in "static" transductions, but not in spinoculation, perform similarly well and VueLife bags performed slightly less than Lifecell® X-fold™ bags with respect to transduction efficiency, lymphocyte subset composition, and lymphocyte function. However, all bag brands performed less than Lifecell® X-fold™ cell culture bags in terms of cell yield. Awaiting availability of other suitable alternatives, the CE-marked Cell Differentiation bag can be used for the clinical-scale GMP manufacture of gene-modified immune cells and compensate for the lower cell yield by adjusting numbers of cells at the start of transduction.[134] Hollyman *et al.*[34] applied Permalife PL240 bags (Origen Biomedical, Austin TX) for retronectin-assisted retroviral transduction, in combination with spinoculation for 1 h at room temperature, at 186 g using a HL-2B rotor in a Sorval RC-3BP centrifuge. Alternatively, different transduction platforms may be explored, e.g., Retronectin coated to beads, which eliminates special requirements to the bags used for transduction.[140]

5.3.6.3 *Clinical-scale transfection*

First trials have entered the clinic to test the safety and feasibility of electroporation-based T-cell modification. In a Sleeping Beauty (SB) transposon-mediated engineering of human primary T cells for therapy of CD19+ lymphoid malignancies, CD3 bead-purified human T cells were transfected with SB-CD19 CAR transposon and SB10 transposase plasmids using a Nucleofector device (Program U14) with the human T-cell nucleofector kit (Amaxa, Gaithersburg, MD).[141] For preparation of CD19 CAR gene-modified cytotoxic T lymphocytes for autologous cellular therapy for lymphoma[55]: anti-CD3 mAb activated PBMC were cultured at day 3 and electroporated with linearized CD19R HyTk-pMG using the BioRad Multiporator® (BioRad Laboratories, Hercules, CA, USA). In an adoptive immunotherapy trial for indolent non-Hodgkin lymphoma and mantle cell lymphoma using genetically modified autologous CD20-specific T cells,[54] PBMC obtained by apheresis were cultured overnight with 30 ng anti-CD3 mAb with 50 IU IL-2 and subsequently electroporated with linearized plasmid encoding a CD20-specific CAR in chilled 0.2 cm electroporation cuvettes using an Eppendorf Multiporator.[131,132]

Following transfection procedures, T cells are subjected to selection,[54,55] and re-stimulation[50,55] in order to generate sufficient numbers of therapeutic cells.

5.3.7 *Preparation of the final product*

For final product preparation, the T-cell cultures have to be concentrated and washed. The culture size determines whether this can be done by centrifugation using blood bags,[27,31] or using a automatic cell concentration and washer system, such as the Cytomate (Cytomate cell washer (Baxter, Deerfield, IL),[32,34] or by continuous centrifuge cell harvester system, or CS3000 (Baxter, Deerfield, IL).[6,55]

In case, T-cell cultures were activated with paramagnetic CD3/CD28 mAb-coated beads, the cultures have to be de-beaded using, e.g., the Dynal ClinEXVIVO MPC magnet (Invitrogen),[34] Isolex device (Baxter, no longer available),[128] or CliniMACS® (Miltenyi).[142,143]

For cell washing, phosphate buffered saline (PBS/EDTA solution) supplied by Baxter and Miltenyi is to date the only GMP buffered saline; however, it does include EDTA as its intended purpose is for magnetic bead cell separation. Alternatively, clinical infusion fluids can be used; such as saline or ringers lactate supplemented with HSA at a final concentration of 1% HSA. The HSA acts as a ubiquitous protein to coat all surfaces of bags and tubing to reduce the likelihood of cells binding to the plastic and hence reducing cell loss.

Of note, during the last years, the availability of several disposables and devices referred to in this overview has been discontinued. The shutdown of the Cell Therapy division of the Baxter Healthcare Cooperation has forced groups to evaluate new brands of critical disposables, and if not available, to adapt procedures. Especially, the discontinuation of the availability of the Isolex — Cytomate disposables rendered costly apparatuses useless and while the cell therapy industry is still in the initial stages, this will be an ongoing problem.

5.3.8 **Product check**

5.3.8.1 *In process and release assays*

The cell cultures are maintained by regular checking and adjusting the cell concentration and assessing viability. This can be done by hand using a hemocytometer in combination with viability stains such as trypan blue that aids in the identification of both viable and dead cells, or by using automated hemocytometers, e.g., Ac·T 8 hematology analyzer (Beckman Coulter). Flow cytometric analysis is informative for the composition of the T-cell culture; typically the percentage of CD3+ T cells and the proportion of cells expressing the transgene of interest. These proportion might serve as release criteria for the final product, i.e., in the CAIX CAR trial,[32] the final product should contain >90% viable cells, >90% CD3+ T cells, and >20% CAIX CAR expressing CD3+ T cells. In addition, the final product check might further include assessment of non-T cells, as NK cells (CD16+CD3-CD45+), and the T-cell subset distribution, CD4+ vs CD8+ T cells % TCR$\alpha\beta$+ and T-cell differentiation marker distribution.

5.3.8.2 *Biosafety testing*

Process control and release criteria include negative bacterial and fungal cultures of the starting material and of the final product, or of samples taken from culture 48 h prior to preparation of the final product.[32] In addition, negative mycoplasma cultures and endotoxin levels <5 EU/kg might be required for the final product.[34,54]

An important issue is the testing for RCR and individual national RCR-testing policies are generally based on the FDA/CBER guidance — *"Guidance for industry: Supplemental guidance on testing for replication competent retrovirus in retroviral vector based gene therapy products and during follow-up of patients in clinical trials using retroviral vectors"*,[93] defining that besides master/working cell bank and vector batch, also the *ex vivo*-transduced cells, when cultured for at least 4 days, should be tested for RCR, irrespective of the type and nature of the transduced cells. Lamers *et al.*[94] reported that GALV

RCR poorly replicate and do not spread in primary human T-cell cultures. In addition, Ebeling et al.[91] reported that human primary T cells have a low capacity to amplify MLV-based amphotropic RCR and that the virions produced are largely non-infectious. These observations question the relevance of testing ex vivo gene-transduced primary human T lymphocytes for GALV RCR and MLV RCR, in addition to having negative RCR test results for the vector-producing cells and the clinical vector.

The Dutch and the UK regulatory authorities do not require RCR testing of the patient T-cell doses following a successful process validation, in which five clinical transductions have proved to be RCR free. Yet, aliquots of transduced patient T-cell doses are stored for retrospective analyses if clinically indicated.

5.3.8.3 Impurities

Following bead-assisted activation of T cells and removal of these beads from the final product, the final product has to be checked for contamination beads (<100 beads per 3.10^6 T cells).[54]

Non-T cells, and non-transduced cells might be considered as impurities of the therapeutic product. In suicide gene controlled allogeneic donor lymphocyte infusion (DLI) protocols, non-transduced T cells have to be removed from the product for safety reasons (uncontrolled graft versus host disease). However, no such risk exists in autologous use of gene-modified T cells for cancer therapy. Therefore, these T-cell preparations are infused as such, without selection of the transduced cells.

Potential impurities are associated with the RTVsup, antibodies, and cytokines to be carried through the process as well as residual culture medium components. These potential impurities have not been specifically tested for or controlled for in the final product as the possibility of their presence, following the multiple washing and dilution steps, is believed to be nil. This deduction along with the clinical history of ex vivo gene manipulation of T cells dating as far back as 1990 with ADA-SCID patients[4,76,144-146] has convinced regulatory bodies in Europe and the US that this is a suitably low risk for the category of patients who are eligible for CAR- and TCR-related trials (i.e., terminally ill cancer patients).

5.3.8.4 Gene modification-induced T-cell transformation

One of the potential drawbacks of viral-assisted gene modification is the occurrence of putative recombination events leading to cell transformation and thus generating malignant T cells. These cells will grow independent of exogenous IL-2; assessment of IL-2 dependent growth of the therapeutic cells documents safety of the product in this respect.[54,131] In addition, spectra-type analysis of the final product may also be informative in this context, e.g., showing a normal versus a skewed repertoire or skewed single Vb family.[34]

5.3.8.5 Functional testing

The gene-modified T cells are tested for transgene-specific functions, such as *in vitro* cytotoxicity, or cytokine release, or gene upregulation when cultured in presence of immobilized target antigen or antigen expressing cells. These activities are usually documented extensively in the preclinical phase of the project, and are often not mandatory prior to release of the therapeutic cells, but are documented retrospectively. In specific applications, one might require functional assessment of each individual patient lot prior to infusion.

5.3.9 Preparation of clinical infusion product

Following concentration and wash of the final product and check of the release criteria, the therapeutic cells can be prepared for direct clinical infusion. Cells are resuspended in an infusion fluid, such as isotonic saline,[55] Plasmalyte A (Baxter),[34] or Ringers lactate[32] and preferably supplemented with at least 1% HSA.

Alternatively, therapeutic cells might need to be frozen in order to perform further product characterization and safety testing or patient preparation, using as cryopreservation medium, e.g., Plasmalyte

A supplemented with 5% HSA and 10% DMSO[54,127] and the product stored in the liquid nitrogen, vapor phase until required.

At the day of infusion, cells are thawed and infused without washing, as is common practice in peripheral blood stem cell donations, or alternatively DMSO is washed out using a Cytomate cell washer and incubated for 3–4 h prior infusion.[54]

5.4 Data Management and Regulatory Aspects

5.4.1 Data management in the cell therapy production facility

The activities of cell therapy facilities are associated with substantial amounts of information. For reasons of best practice, regulation and adherence to prevailing standards, the data generated in the course of cell therapy product processing must be recorded and retained in an organized manner.

Because cell therapy products are functionally pharmaceuticals, the paradigm of the pharmaceutical manufacturing batch process record (BPR) is proposed as a unit for collecting the data resulting from processing.[147]

5.4.2 Regulatory aspects

Clinical trials using engineered T cells meet strict regulations and need several approvals and permissions. These aspects will only be discussed briefly, as they rely on national law defined rules and regulations.

In the USA, *ex vivo*-manipulated cell products for infusion must undergo review as an IND with the FDA. This requires successful filing of an IND application, institutional review board (IRB) approval of the protocol and consent, and is governed by the standards set out in the Code of Federal Regulations (CFR) sections 21CFR 210, 211 (Good Manufacturing Practices or GMP), 21 CFR 1270 Human Tissue Intended for Transplantation and 21 CFR1271 Human Cells, Tissues, and Cellular and Tissue-Based Products (collectively referred to as Good Tissue Practices or GTP). GMP regulations dictate the requirement for a quality-systems approach to T-cell manufacturing that includes control over manufacturing, personnel, facilities, raw materials compounding, and final product release, as well as a complete documentation system for production, testing, and product release. When gene transfer is proposed as part of the manufacturing process, specific additional approval is required from an institutional biosafety committee (IBC), which reviews compliance with reporting requirements defined in the FDA's "Points To Consider in the Design and Submission of Protocols For The Transfer of Recombinant DNA Molecules Into One or More Human Research Participants", also known as "Appendix M", and by the National Institutes of Health Office of Biotechnology Activities (NIH-OBA) and NIH Recombinant DNA Advisory Committee (RAC), see Ref. 33.

In Europe, a clinical viral vector has to be approved by the national competent authority based on an environmental risk assessment according to Directive 2001/18/EC. According the EU regulations, gene-modified T-cell products fit the description of advanced therapy medicinal products (ATMP; Regulation no 1394/2007 on advanced therapy medicinal products and amending Directive 2001/83/EC and Regulation (EC) No 726/2004). However, this regulation is set up to provide a mechanism in which the European Medicine Agency can give marketing authorization to companies. As long as the ATMP is part of clinical study, no EMA authorization is required, and the clinical trials directive (2001/20/EC; implemented in national law from January 2006) has to be followed, which implies that the clinical trial should be conducted following Good Clinical trial Procedures (GCP). In general, GCP include that the production of the therapeutics (engineered T cells) is done under conditions of GMP and testing of these products and monitoring of patients is executed in certified laboratories using validated methods.

For gene-modified cell products for infusion, in general, following permits and approvals will be needed and have to be obtained from the national and local competent authorities: (i) permit for use of the vector; (ii) certification of the vector production facility; (iii) approval/release of the clinical virus batch; (iv) certification of the cell processing facility; (v) approval of methodology to produce the gene-modified T cells; and (vi) approval of the clinical protocol.

A clinical vector batch has to meet defined quality and safety criteria, as example, see Table 5.1 for the quality control and safety tests performed on the clinical lot of Phoenix-A clone 58 (SFG-CAIX-CAR) RTVsup and clinical lots of transduced patient T lymphocytes in preparation of the Rotterdam RCC trial. The final gene-modified T-cell product will have to meet the defined release criteria, and will be released for clinical application by a qualified person (gene therapy), in general a pharmacist.

5.5 Future Developments

Developments in the gene transfer technologies to achieve optimal engineered T cells for cancer therapy have been discussed on previous chapters. Yet, the outcome of these evaluations, whether we continue with optimized SIN γ-retro- or lentivirus vectors, whether or not to include site-specific integration using Zn fingers, or change to foamy virus vectors, or non-viral gene transfer, using for instance the transposon system, will finally determine the technological requirements for generation of therapeutic numbers of the gene-modified T cells. Gene transfer at clinical scale has been optimized for γ-retro- or lentivirus vectors, and are under development for the non-viral transfer technologies, yet there is still room for all protocols to improve in context of improving the manufacturing process, reducing labor and cost.

In the search for the most optimal cells for adoptive immunotherapy, it has become clear that the focus has to be moved from effector (memory) T cells (T_{EM}) to central memory T cells (T_{CM}) or cells within the T_{CM} that have retained their "stemness" and have referred to as T memory stem cells (T_{SCM} cells).[148] Generating T_{SCM} cells means that fewer cells may be required for adoptive immunotherapy, making a complex technique less cumbersome and potentially more widely used. In the present engineered T-cell protocols, the T cells need to be activated and expanded to generate therapeutic numbers, which drives these cells into less desirable maturation stages. Recent developments show ways to achieve gene-modified T cells at less matured stages, such as: (i) gene transfer technologies that do not need vigorous activation of the starting T-cell population, such as non-viral or lentiviral vectors; (ii) T-cell culture conditions, including other members of γ-chain family of cytokines, like IL-7, IL-15, and IL-21, that induced minimal differentiation and maximal antigen-specificity. GMP-grade preparations of IL-7, IL-15, and IL-21 are available or will become available shortly for clinical evaluation.

In this respect, it is also interesting to mention that Gattinoni et al.[149,150] reported that activating the Wnt-β-catenin pathway in T cells, e.g., by inhibition of glycogen synthase kinase-3β (Gsk-3β), during the initial priming step, arrests effector T-cell differentiation and generates CD8$^+$ memory stem cells, that maintain this property following further processing.

The classical T-cell activation by soluble CD3 mAb (OKT3) is being replaced by activation by the CD3 and CD28 mAbs, in order to generate a full and long-lasting T-cell effector function of the engineered T cells. At present, the two mAb activation method is applied through either beads or as soluble mAbs in case accessory cells such as monocytes are present in the starting population, depending on the laboratory preference, and/or cont(r)act with suppliers.

The generation of engineered T cells is a laborious and costly business and relies on the availability of clinical-grade components, disposables, and appropriate instrumentation and facilities. Procedures and methods are under permanent development and evaluation by changing availabilities of these critical components. With extension of the applications of engineered T cells, the need for a more "industrial" manufacturing process emerged. The development of the high cell density (Wave) culture technology might fill this need. In addition, speeding up the generation of the engineered T cells favors both T-cell quality and reduces cost. In this context, the application of the cellular aAPC is a promising approach, which is presently under clinical evaluation.

At present, there is not one, but there are various "standard" protocols for the generation of therapeutic engineered T cells, as there are also different chimeric receptor formats, and patient treatment schedules. There is an urgent need for standardization in generation and clinical application of receptor-modified T cells.[151]

References

1. Rosenberg, S.A. *et al.* (1993). Prospective randomized trial of high-dose inter-leukin-2 alone or in conjunction with lymphokine-activated killer cells for the treatment of patients with advanced cancer. *J Natl Cancer Inst* 85, 622–632.
2. Law, T.M. *et al.* (1995). Phase III randomized trial of interleukin-2 with or without lymphokine-activated killer cells in the treatment of patients with advanced renal cell carcinoma. *Cancer* 76, 824–832.
3. Riddell, S.R. *et al.* (1992). Restoration of viral immunity in immunodeficient humans by the adoptive transfer of T cell clones. *Science* 257, 238–241.
4. Rooney, C.M. *et al.* (1998). Infusion of cytotoxic T cells for the prevention and treatment of Epstein-Barr virus-induced lymphoma in allogeneic transplant recipients. *Blood* 92, 1549–1555.
5. Rosenberg, S.A. *et al.* (1994). Treatment of patients with metastatic melanoma with autologous tumor-infiltrating lymphocytes and interleukin 2. *J Natl Cancer Inst* 86, 1159–1166.
6. Dudley, M.E. *et al.* (2002). A phase I study of nonmyeloablative chemotherapy and adoptive transfer of autologous tumor antigen-specific T lymphocytes in patients with metastatic melanoma. *J Immunother* 25, 243–251.
7. Yee, C. *et al.* (2002). Adoptive T cell therapy using antigen-specific CD8+ T cell clones for the treatment of patients with metastatic melanoma: In vivo persistence, migration, and antitumor effect of transferred T cells. *Proc Natl Acad Sci USA* 99, 16168–16173.
8. Dudley, M.E. *et al.* (2008). Adoptive cell therapy for patients with metastatic melanoma: Evaluation of intensive myeloablative chemoradiation preparative regimens. *J Clin Oncol* 26, 5233–5239.
9. Dudley, M.E. *et al.* (2005). Adoptive cell transfer therapy following non-myeloablative but lymphodepleting chemotherapy for the treatment of patients with refractory metastatic melanoma. *J Clin Oncol* 23, 2346–2357.
10. Rosenberg, S.A. and Dudley, M.E. (2009). Adoptive cell therapy for the treat-ment of patients with metastatic melanoma. *Curr Opin Immunol* 21, 233–240.
11. Willemsen, R.A., Debets, R., Chames, P., and Bolhuis, R.L. (2003). Genetic engineering of T cell specificity for immunotherapy of cancer. *Hum Immunol* 64, 56–68.
12. Koelemij, R. *et al.* (1999). Bispecific antibodies in cancer therapy, from the laboratory to the clinic. *J Immunother* 22, 514–524.
13. Bolhuis, R.L. *et al.* (1992). Adoptive immunotherapy of ovarian carcinoma with bs-MAb-targeted lymphocytes: A multicenter study. *Int J Cancer Suppl* 7, 78–81.
14. Canevari, S. *et al.* (1995). Bispecific antibody targeted T cell therapy of ovarian cancer: Clinical results and future directions. *J Hematother* 4, 423–427.

15. Blank-Voorthuis, C.J. *et al.* (1993). Clustered CD3/TCR complexes do not transduce activation signals after bispecific monoclonal antibody-triggered lysis by cytotoxic T lymphocytes via CD3. *J Immunol* 151, 2904–2914.

16. Lamers, C.H., Gratama, J.W., Warnaar, S.O., Stoter, G., and Bolhuis, R.L. (1995). Inhibition of bispecific monoclonal antibody (bsAb)-targeted cytolysis by human anti-mouse antibodies in ovarian carcinoma patients treated with bsAb-targeted activated T-lymphocytes. *Int J Cancer* 60, 450–457.

17. Lamers, C.H., Bolhuis, R.L., Warnaar, S.O., Stoter, G., and Gratama, J.W. (1997). Local but no systemic immunomodulation by intraperitoneal treatment of advanced ovarian cancer with autologous T lymphocytes re-targeted by a bi-specific monoclonal antibody. *Int J Cancer* 73, 211–219.

18. Willemsen, R.A. *et al.* (2000). Grafting primary human T lymphocytes with cancer-specific chimeric single chain and two chain TCR. *Gene Ther* 7, 1369–1377.

19. Eshhar, Z. (1997). Tumor-specific T-bodies: Towards clinical application. *Cancer Immunol Immunother* 45, 131–136.

20. Weijtens, M.E., Willemsen, R.A., Valerio, D., Stam, K., and Bolhuis, R.L. (1996). Single chain Ig/gamma gene-redirected human T lymphocytes produce cytokines, specifically lyse tumor cells, and recycle lytic capacity. *J Immunol* 157, 836–843.

21. Eshhar, Z., Waks, T., Gross, G., and Schindler, D.G. (1993). Specific activation and targeting of cytotoxic lymphocytes through chimeric single chains consisting of antibody-binding domains and the gamma or zeta subunits of the immunoglobulin and T-cell receptors. *Proc Natl Acad Sci USA* 90, 720–724.

22. Sadelain, M., Brentjens, R., and Riviere, I. (2009). The promise and potential pitfalls of chimeric antigen receptors. *Curr Opin Immunol* 21, 215–223.

23. Schumacher, T.N. and Restifo, N.P. (2009). Adoptive T cell therapy of cancer. *Curr Opin Immunol* 21, 187–189.

24. Muul, L.M., Director, E.P., Hyatt, C.L. and Rosenberg, S.A. (1986). Large scale production of human lymphokine activated killer cells for use in adoptive immunotherapy. *J Immunol Methods* 88, 265–275.

25. Muul, L.M. *et al.* (1987). Studies of serum-free culture medium in the generation of lymphokine activated killer cells. *J Immunol Methods* 105, 183–192.

26. Yee, C., Savage, P.A., Lee, P.P., Davis, M.M., and Greenberg, P.D. (1999). Isolation of high avidity melanoma-reactive CTL from heterogeneous populations using peptide-MHC tetramers. *J Immunol* 162, 2227–2234.

27. Lamers, C.H. *et al.* (1992). Optimization of culture conditions for activation and large-scale expansion of human T lymphocytes for bispecific antibody-directed cellular immunotherapy. *Int J Cancer* 51, 973–979.

28. Topalian, S.L., Muul, L.M., Solomon, D., and Rosenberg, S.A. (1987). Expansion of human tumor infiltrating lymphocytes for use in immunotherapy trials. *J Immunol Methods* 102, 127–141.

29. Dudley, M.E., Wunderlich, J.R., Shelton, T.E., Even, J., and Rosenberg, S.A. (2003). Generation of tumor-infiltrating lymphocyte cultures for use in adoptive transfer therapy for melanoma patients. *J Immunother* 26, 332–342.

30. Besser, M.J. *et al.* (2009). Minimally cultured or selected autologous tumor-infiltrating lymphocytes after a lympho-depleting chemotherapy regimen in metastatic melanoma patients. *J Immunother* 32, 415–423.

31. Lamers, C.H., Willemsen, R.A., Luider, B.A., Debets, R., and Bolhuis, R.L. (2002). Protocol for gene transduction and expansion of human T lymphocytes for clinical immunogene therapy of cancer. *Cancer Gene Ther* 9, 613–623.

32. Lamers, C., van Elzakker, P., Langeveld, S., Sleijfer, S., and Gratama, J. (2006). Process validation and clinical evaluation of a protocol to generate gene-modified T lymphocytes for imunogene therapy for metastatic renal cell carcinoma: GMP-controlled transduction and expansion of patient's T lymphocytes using a carboxy anhydrase IX-specific scFv transgene. *Cytotherapy* 8, 542–553.

33. DiGiusto, D.L. and Cooper, L.J. (2007). Preparing clinical grade Ag-specific T cells for adoptive immunotherapy trials. *Cytotherapy* 9, 613–629.

34. Hollyman, D. *et al.* (2009). Manufacturing validation of biologically functional T cells targeted to CD19 antigen for autologous adoptive cell therapy. *J Immunother* 32, 169–180.

35. Muul, L.M. *et al.* (1987). Development of an automated closed system for generation of human lymphokine-activated killer (LAK) cells for use in adoptive immunotherapy. *J Immunol Methods* 101, 171–181.

36. Robinet, E. *et al.* (1998). A closed culture system for the ex vivo transduction and expansion of human T lymphocytes. *J Hematother* 7, 205–215.

37. Turin, I. *et al.* (2007). GMP production of anti-tumor cytotoxic T-cell lines for adoptive T-cell therapy in patients with solid neoplasia. *Cytotherapy* 9, 499–507.

38. Tran, C.A. *et al.* (2007). Manufacturing of large numbers of patient-specific T cells for adoptive immunotherapy: An approach to improving product safety, composition, and production capacity. *J Immunother* 30, 644–654.

39. Sili, U. *et al.* (2012). Production of good manufacturing practice-grade cytotoxic T lymphocytes specific for Epstein-Barr virus, cytomegalovirus and adenovirus to prevent or treat viral infections post-allogeneic hematopoietic stem cell transplant. *Cytotherapy* 14, 7–11.

40. Grilley, B.J. and Gee, A.P. (2003). Gene transfer: Regulatory issues and their impact on the clinical investigator and the good manufacturing production facility. *Cytotherapy* 5, 197–207.

41. Burger, S.R. (2003). Current regulatory issues in cell and tissue therapy. *Cytotherapy* 5, 289–298.

42. D'Costa, J., Mansfield, S.G., and Humeau, L.M. (2009). Lentiviral vectors in clinical trials: Current status. *Curr Opin Mol Ther* 11, 554–564.

43. Naldini, L. *et al.* (1996). In vivo gene delivery and stable transduction of non-dividing cells by a lentiviral vector. *Science* 272, 263–267.

44. Amado, R.G. and Chen, I.S. (1999). Lentiviral vectors — The promise of gene therapy within reach? *Science* 285, 674–676.

45. Ellis, J. (2005). Silencing and variegation of gammaretrovirus and lentivirus vectors. *Hum Gene Ther* 16, 1241–1246.

46. Levine, B.L. *et al.* (2006). Gene transfer in humans using a conditionally replicating lentiviral vector. *Proc Natl Acad Sci USA* 103, 17372–17377.

47. Russell, D.W. and Miller, A.D. (1996). Foamy virus vectors. *J Virol* 70, 217–222.

48. Beard, B.C. *et al.* (2007). Unique integration profiles in a canine model of long-term repopulating cells transduced with gammaretrovirus, lentivirus, or foamy virus. *Hum Gene Ther* 18, 423–434.

49. Izsvak, Z., Chuah, M.K., Vandendriessche, T., and Ivics, Z. (2009). Efficient stable gene transfer into human cells by the Sleeping Beauty transposon vectors. *Methods* 49, 287–297.

50. Huang, X. *et al.* (2008). Sleeping Beauty transposon-mediated engineering of human primary T cells for therapy of CD19+ lymphoid malignancies. *Mol Ther* 16, 580–589.

51. Singh, H. *et al.* (2008). Redirecting specificity of T-cell populations for CD19 using the Sleeping Beauty system. *Cancer Res* 68, 2961–2971.

52. Mitchell, D.A. *et al.* (2008). Selective modification of antigen-specific T cells by RNA electroporation. *Hum Gene Ther* 19, 511–521.

53. Rowley, J., Monie, A., Hung, C.F., and Wu, T.C. (2009). Expression of IL-15RA or an IL-15/IL-15RA fusion on CD8+ T cells modifies adoptively transferred T-cell function in cis. *Eur J Immunol* 39, 491–506.

54. Till, B.G. *et al.* (2008). Adoptive immunotherapy for indolent non-Hodgkin lymphoma and mantle cell lymphoma using genetically modified autologous CD20-specific T cells. *Blood* 112, 2261–2271.

55. Cooper, L.J. *et al.* (2006). Manufacturing of gene-modified cytotoxic T lymphocytes for autologous cellular therapy for lymphoma. *Cytotherapy* 8, 105–117.

56. Jaffee, E.M. *et al.* (1993). High efficiency gene transfer into primary human tumor explants without cell selection. *Cancer Res* 53, 2221–2226.

57. Riviere, I., Brose, K., and Mulligan, R.C. (1995). Effects of retroviral vector design on expression of human adenosine deaminase in murine bone marrow transplant recipients engrafted with genetically modified cells. *Proc Natl Acad Sci USA* 92, 6733–6737.

58. Bueler, H. and Mulligan, R.C. (1996). Induction of antigen-specific tumor immunity by genetic and cellular vaccines against MAGE: Enhanced tumor protection by coexpression of granulocyte-macrophage colony-stimulating factor and B7-1. *Mol Med* 2, 545–555.

59. Kershaw, M.H. *et al.* (2006). A phase I study on adoptive immunotherapy using gene-modified T cells for ovarian cancer. *Clin Cancer Res* 12, 6106–6115.

60. Lamers, C.H. *et al.* (2006). Treatment of metastatic renal cell carcinoma with autologous T-lymphocytes genetically retargeted against carbonic anhydrase IX: First clinical experience. *J Clin Oncol* 24, e20–e22.

61. Morgan, R.A. *et al.* (2006). Cancer regression in patients after transfer of genetically engineered lymphocytes. *Science* 314, 126–129.

62. Johnson, L.A. *et al.* (2009). Gene therapy with human and mouse T-cell receptors mediates cancer regression and targets normal tissues expressing cognate antigen. *Blood* 114, 535–546.

63. Schambach, A. *et al.* (2000). Context dependence of different modules for posttranscriptional enhancement of gene expression from retroviral vectors. *Mol Ther* 2, 435–445.

64. Engels, B. *et al.* (2003). Retroviral vectors for high-level transgene expression in T lymphocytes. *Hum Gene Ther* 14, 1155–1168.

65. Loew, R. *et al.* (2010). A new PG13-based packaging cell line for stable production of clinical-grade self-inactivating gamma-retroviral vectors using targeted integration. *Gene Ther* 17, 272–280.

66. Uckert, W. and Schumacher, T.N. (2009). TCR transgenes and transgene cassettes for TCR gene therapy: Status in 2008. *Cancer Immunol Immunother* 58, 809 822.

67. Leisegang, M. *et al.* (2008). Enhanced functionality of T cell receptor-redirected T cells is defined by the transgene cassette. *J Mol Med* 86, 573–583.

68. Bauer, Jr., T.R., Miller, A.D., and Hickstein, D.D. (1995). Improved transfer of the leukocyte integrin CD18 subunit into hematopoietic cell lines by using retroviral vectors having a gibbon ape leukemia virus envelope. *Blood* 86, 2379–2387.

69. Uckert, W. *et al.* (2000). Efficient gene transfer into primary human CD8+ T lymphocytes by MuLV-10A1 retrovirus pseudotype. *Hum Gene Ther* 11, 1005–1014.

70. Cronin, J., Zhang, X.Y., and Reiser, J. (2005). Altering the tropism of lentiviral vectors through pseudotyping. *Current Gene Ther* 5, 387–398.

71. Bunnell, B.A., Muul, L.M., Donahue, R.E., Blaese, R.M., and Morgan, R.A. (1995). High-efficiency retroviral-mediated gene transfer into human and non-human primate peripheral blood lymphocytes. *Proc Natl Acad Sci USA* 92, 7739–7743.

72. Lam, J.S., Reeves, M.E., Cowherd, R., Rosenberg, S.A., and Hwu, P. (1996). Improved gene transfer into human lymphocytes using retroviruses with the gibbon ape leukemia virus envelope. *Hum Gene Ther* 7, 1415–1422.

73. Miller, A.D. and Buttimore, C. (1986). Redesign of retrovirus packaging cell lines to avoid recombination leading to helper virus production. *Mol Cell Biol* 6, 2895–2902.

74. Markowitz, D., Goff, S., and Bank, A. (1988). Construction and use of a safe and efficient amphotropic packaging cell line. *Virology* 167, 400–406.

75. Rosenberg, S.A. *et al.* (1990). Gene transfer into humans — Immunotherapy of patients with advanced melanoma, using tumor-infiltrating lymphocytes modified by retroviral gene transduction. *N Engl J Med* 323, 570–578.

76. Bordignon, C. *et al.* (1995). Gene therapy in peripheral blood lymphocytes and bone marrow for ADA- immunodeficient patients. *Science* 270, 470–475.

77. Verzeletti, S. *et al.* (1998). Herpes simplex virus thymidine kinase gene transfer for controlled graft-versus-host disease and graft-versus-leukemia: Clinical follow-up and improved new vectors. *Hum Gene Ther* 9, 2243–2251.

78. Miller, A.D. *et al.* (1991). Construction and properties of retrovirus packaging cells based on gibbon ape leukemia virus. *J Virol* 65, 2220–2224.

79. Swift, S., Lorens, J., Achacoso, P., and Nolan, G.P. (1999). Radid production of retroviruses for efficient gene delivery to mammalian cells using 293T cell-basd systems. *Curr Protoc Immunol* 10.17.14–10.17.29.

80. van der Loo, J.C. *et al.* (2012). Scale-up and manufacturing of clinical-grade self-inactivating gamma-retroviral vectors by transient transfection. *Gene Ther* 19, 246–254.

81. Ghani, K. *et al.* (2009). Efficient human hematopoietic cell transduction using RD114- and GALV-pseudotyped retroviral vectors produced in suspension and serum-free media. *Hum Gene Ther* 20, 966–974.

82. Finer, M.H., Dull, T.J., Qin, L., Farson, D., and Roberts, M.R. (1994). Kat: A high-efficiency retroviral transduction system for primary human T lymphocytes. *Blood* 83, 43–50.

83. Lamers, C.H. *et al.* (2006). Phoenix-ampho outperforms PG13 as retroviral packaging cells to transduce human T cells with tumor-specific receptors: Implications for clinical immunogene therapy of cancer. *Cancer Gene Ther* 13, 503–509.

84. Cornetta, K., Matheson, L., and Ballas, C. (2005). Retroviral vector production in the National Gene Vector Laboratory at Indiana University. *Gene Ther* 12(Suppl 1), S28–S35.

85. Lamers, C.H. *et al.* (2008). Retroviral vectors for clinical immunogene therapy are stable for up to 9 years. *Cancer Gene Ther* 15, 268–274.

86. Lander, M.R. and Chattopadhyay, S.K. (1984). A Mus dunni cell line that lacks sequences closely related to endogenous murine leukemia viruses and can be infected by ectropic, amphotropic, xenotropic, and mink cell focus-forming viruses. *J Virol* 52, 695–698.

87. Gunter, K.C., Khan, A.S., and Noguchi, P.D. (1993). The safety of retroviral vectors. *Hum Gene Ther* 4, 643–645.

88. Wilson, C.A., Ng, T.H., and Miller, A.E. (1997). Evaluation of recommendations for replication-competent retrovirus testing associated with use of retroviral vectors. *Hum Gene Ther* 8, 869–874.

89. Chen, J., Reeves, L., and Cornetta, K. (2001). Safety testing for replication-competent retrovirus associated with gibbon ape leukemia virus-pseudotyped retroviral vectors. *Hum Gene Ther* 12, 61–70.

90. Martineau, D. *et al.* (1997). Evaluation of PCR and ELISA assays for screening clinical trial subjects for replication-competent retrovirus. *Hum Gene Ther* 8, 1231–1241.

91. Ebeling, S.B. *et al.* (2003). Human primary T lymphocytes have a low capacity to amplify MLV-based amphotropic RCR and the virions produced are largely noninfectious. *Gene Ther* 10, 1800–1806.

92. Sastry, L. *et al.* (2005). Product-enhanced reverse transcriptase assay for replication-competent retrovirus and lentivirus detection. *Hum Gene Ther* 16, 1227–1236.

93. Guidance for Industry, U.S.Department of Health and Human Services, Food and Drug Administration, and Center for Biologics Evalution and Research (CBER) (2001). Supplemental guidance on testing for replication-competent retrovirus in retroviral vector-based gene therapy products and during follow-up of patients in clinical trials using retroviral vectors. *Hum Gene Ther* 12, 315–320.

94. Lamers, C.H., Willemsen, R.A., van Elzakker, P.M., Gratama, J.W., and Debets, R. (2009). Gibbon ape leukemia virus poorly replicates in primary human T lymphocytes: Implications for safety testing of primary human T lymphocytes transduced with GALV pseudotyped vectors. *J Immunother* 32, 272–279.

95. Zufferey, R. *et al.* (1998). Self-inactivating lentivirus vector for safe and efficient in vivo gene delivery. *J Virol* 72, 9873–9880.

96. Miyoshi, H., Blomer, U., Takahashi, M., Gage, F.H., and Verma, I.M. (1998). Development of a self-inactivating lentivirus vector. *J Virol* 72, 8150–8157.

97. Pluta, K. and Kacprzak, M.M. (2009). Use of HIV as a gene transfer vector. *Acta Biochim Pol* 56, 531–595.

98. Dissen, G.A. *et al.* (2009). In vivo manipulation of gene expression in non-human primates using lentiviral vectors as delivery vehicles. *Methods* 49, 70–77.

99. Ni, Y. *et al.* (2005). Generation of a packaging cell line for prolonged large-scale production of high-titer HIV-1-based lentiviral vector. *J Gene Med* 7, 818–834.

100. Broussau, S. *et al.* (2008). Inducible packaging cells for large-scale production of lentiviral vectors in serum-free suspension culture. *Mol Ther* 16, 500–507.

101. Ansorge, S. *et al.* (2009). Development of a scalable process for high-yield lentiviral vector production by transient transfection of HEK293 suspension cultures. *J Gene Med* 11, 868–876.

102. Cornetta, K. *et al.* (2011). Replication-competent lentivirus analysis of clinical grade vector products. *Mol Ther* 19, 557–566.

103. Schonely, K. *et al.* (2003). QC release testing of an HIV-1 based lentiviral vector lot and transduced cellular product. *www. bioprocessingjournal.com* July/August, 39–47.

104. Manilla, P. *et al.* (2005). Regulatory considerations for novel gene therapy products: A review of the process leading to the first clinical lentiviral vector. *Hum Gene Ther* 16, 17–25.

105. Cartier, N. *et al.* (2009). Hematopoietic stem cell gene therapy with a lentiviral vector in X-linked adrenoleukodystrophy. *Science* 326, 818–823.

106. Sadelain, M. *et al.* (2010). Strategy for a multicenter phase I clinical trial to evaluate globin gene transfer in beta-thalassemia. *Ann N Y Acad Sci* 1202, 52–58.

107. Porter, D.L., Levine, B.L., Kalos, M., Bagg, A., and June, C.H. (2011). Chimeric antigen receptor-modified T cells in chronic lymphoid leukemia. *N Engl J Med* 365, 725–733.

108. Kalos, M. *et al.* (2011). T cells with chimeric antigen receptors have potent antitumor effects and can establish memory in patients with advanced Leukemia. *Sci Transl Med* 3, 95ra73.

109. Coccoris, M. *et al.* (2010). T cell receptor (TCR) gene therapy to treat melanoma: Lessons from clinical and preclinical studies. *Expert Opin Biol Ther* 10, 547–562.

110. Brosnan, J.T. (2003). Interorgan amino acid transport and its regulation. *J Nutr* 133, 2068S–2072S.

111. Carlens, S. *et al.* (2000). Ex vivo T lymphocyte expansion for retroviral transduction: Influence of serum-free media on variations in cell expansion rates and lymphocyte subset distribution. *Exp Hematol* 28, 1137–1146.

112. Block, A. *et al.* (2008). Impact of cell culture media on the expansion efficiency and T-cell receptor Vbeta (TRBV) repertoire of in vitro expanded T cells using feeder cells. *Med Sci Monit* 14, BR88–BR95.

113. Miller, D.G., Adam, M.A., and Miller, A.D. (1990). Gene transfer by retrovirus vectors occurs only in cells that are actively replicating at the time of infection. *Mol Cell Biol* 10, 4239–4242.

114. Sauce, D. *et al.* (2002). Retrovirus-mediated gene transfer in primary T lymphocytes impairs their anti-Epstein-Barr virus potential through both culture-dependent and selection process-dependent mechanisms. *Blood* 99, 1165–1173.

115. Bondanza, A. *et al.* (2006). Suicide gene therapy of graft-versus-host disease induced by central memory human T lymphocytes. *Blood* 107, 1828–1836.

116. Levine, B.L. (2008). T lymphocyte engineering ex vivo for cancer and infectious disease. *Expert Opin Biol Ther* 8, 475–489.

117. Uberti, J.P. *et al.* (1994). Preclinical studies using immobilized OKT3 to activate human T cells for adoptive immunotherapy: Optimal conditions for the proliferation and induction of non-MHC-restricted cytotoxicity. *Clin Immunol Immunopathol* 70, 234–240.

118. Rosenberg, S.A. *et al.* (1994). Treatment of 283 consecutive patients with metastatic melanoma or renal cell cancer using high-dose bolus interleukin 2. *JAMA* 271, 907–913.

119. Besser, M.J. *et al.* (2009). Modifying interleukin-2 concentrations during culture improves function of T cells for adoptive immunotherapy. *Cytotherapy* 11, 206–217.

120. Leonard, W.J., Zeng, R., and Spolski, R. (2008). Interleukin 21: A cytokine/cytokine receptor system that has come of age. *J Leukoc Biol* 84, 348–356.

121. Pouw, N., Treffers-Westerlaken, E., Mondino, A., Lamers, C., and Debets, R. (2010). TCR gene-engineered T cell: Limited T cell activation and combined use of IL-15 and IL-21 ensure minimal differentiation and maximal antigen-specificity. *Mol Immunol.*

122. Pollok, K.E. *et al.* (1998). High-efficiency gene transfer into normal and adenosine deaminase-deficient T lymphocytes is mediated by transduction on recombinant fibronectin fragments. *J Virol* 72, 4882–4892.

123. Hanenberg, H. *et al.* (1996). Colocalization of retrovirus and target cells on specific fibronectin fragments increases genetic transduction of mammalian cells. *Nat Med* 2, 876–882.

124. Jayasinghe, S.M. *et al.* (2006). Sterile and disposable fluidic subsystem suitable for clinical high speed fluorescence-activated cell sorting. *Cytometry B Clin Cytom* 70, 344–354.

125. Suhoski, M.M. *et al.* (2007). Engineering artificial antigen-presenting cells to express a diverse array of co-stimulatory molecules. *Mol Ther* 15, 981–988.

126. Powell, Jr., D.J. *et al.* (2009). Efficient clinical-scale enrichment of lymphocytes for use in adoptive immunotherapy using a modified counterflow centrifugal elutriation program. *Cytotherapy* 11, 923–935.

127. Pule, M.A. *et al.* (2008). Virus-specific T cells engineered to coexpress tumor-specific receptors: Persistence and antitumor activity in individuals with neuroblastoma. *Nat Med* 14, 1264–1270.

128. Orchard, P.J. *et al.* (2002). Clinical-scale selection of anti-CD3/CD28-activated T cells after transduction with a retroviral vector expressing herpes simplex virus thymidine kinase and truncated nerve growth factor receptor. *Hum Gene Ther* 13, 979–988.

129. Besser, M.J. *et al.* (2006). Adoptive cell therapy for metastatic melanoma patients: Pre-clinical development at the Sheba Medical Center. *Isr Med Assoc J* 8, 164–168.

130. Riddell, S.R. and Greenberg, P.D. (1997). T cell therapy of human CMV and EBV infection in immunocompromised hosts. *Rev Med Virol* 7, 181–192.

131. Jensen, M.C. *et al.* (2010). Antitransgene rejection responses contribute to attenuated persistence of adoptively transferred CD20/CD19-specific chimeric antigen receptor redirected T cells in humans. *Biol Blood Marrow Transplant* 16, 1245–1256.

132. Wang, J. *et al.* (2004). Cellular immunotherapy for follicular lymphoma using genetically modified CD20-specific CD8+ cytotoxic T lymphocytes. *Mol Ther* 9, 577–586.

133. Fowler, D.H. *et al.* (2006). Phase I clinical trial of costimulated, IL-4 polarized donor CD4+ T cells as augmentation of allogeneic hematopoietic cell transplantation. *Biol Blood Marrow Transplant* 12, 1150–1160.

134. Lamers, C.H. *et al.* (2008). Retronectin-assisted retroviral transduction of primary human T lymphocytes under good manufacturing practice conditions: Tissue culture bag critically determines cell yield. *Cytotherapy* 10, 406–416.

135. Kotani, H. *et al.* (1994). Improved methods of retroviral vector transduction and production for gene therapy. *Hum Gene Ther* 5, 19–28.

136. Cornetta, K. and Anderson, W.F. (1989). Protamine sulfate as an effective alternative to polybrene in retroviral-mediated gene-transfer: Implications for human gene therapy. *J Virol Methods* 23, 187–194.

137. Hanenberg, H. *et al.* (1997). Optimization of fibronectin-assisted retroviral gene transfer into human CD34+ hematopoietic cells. *Hum Gene Ther* 8, 2193–2206.

138. Zhou, P., Lee, J., Moore, P., and Brasky, K.M. (2001). High-efficiency gene transfer into rhesus macaque primary T lymphocytes by combining 32 degrees C centrifugation and CH-296-coated plates: Effect of gene transfer protocol on T cell homing receptor expression. *Hum Gene Ther* 12, 1843–1855.

139. Quintas-Cardama, A. *et al.* (2007). Multifactorial optimization of gammaretroviral gene transfer into human T lymphocytes for clinical application. *Hum Gene Ther* 18, 1253–1260.

140. Heemskerk, B. *et al.* (2010). Microbead-assisted retroviral transduction for clinical application. *Hum Gene Ther*.

141. Huang, X. *et al.* (2006). Stable gene transfer and expression in human primary T cells by the Sleeping Beauty transposon system. *Blood* 107, 483–491.

142. McNiece, I.K., Stoney, G.B., Kern, B.P., and Briddell, R.A. (1998). CD34+ cell selection from frozen cord blood products using the Isolex 300i and CliniMACS CD34 selection devices. *J Hematother* 7, 457–461.

143. Watts, M.J. *et al.* (2002). Variable product purity and functional capacity after CD34 selection: A direct comparison of the CliniMACS (v2.1) and Isolex 300i (v2.5) clinical scale devices. *Br J Haematol* 118, 117–123.

144. Blaese, R.M. *et al.* (1995). T lymphocyte-directed gene therapy for ADA-SCID: Initial trial results after 4 years. *Science* 270, 475–480.

145. Woffendin, C., Ranga, U., Yang, Z., Xu, L., and Nabel, G.J. (1996). Expression of a protective gene-prolongs survival of T cells in human immunodeficiency virus-infected patients. *Proc Natl Acad Sci USA* 93, 2889–2894.

146. Muul, L.M. *et al.* (2003). Persistence and expression of the adenosine deaminase gene for 12 years and immune reaction to gene transfer components:

Long-term results of the first clinical gene therapy trial. *Blood* 101, 2563–2569.

147. Janssen, W. (2008). Data management in the cell therapy production facility: The batch process record (BPR). *Cytotherapy* 10, 227–237.

148. Stemberger, C. *et al.* (2009). Stem cell-like plasticity of naive and distinct memory CD8+ T cell subsets. *Semin Immunol* 21, 62–68.

149. Gattinoni, L. *et al.* (2009). Wnt signaling arrests effector T cell differentiation and generates CD8+ memory stem cells. *Nat Med* 15, 808–813.

150. Gattinoni, L. *et al.* (2011). A human memory T cell subset with stem cell-like properties. *Nat Med* 17, 1290–1297.

151. Hawkins, R.E. *et al.* (2010). Development of adoptive cell therapy for cancer: A clinical perspective. *Hum Gene Ther* 21, 665–672.

Chapter 6

Clinical Trial Design

Robert Hawkins, John Haanen and Fiona Thistlethwaite

6.1 Introduction

Successful adoptive cell therapy (ACT) for natural or genetically engineered T cells requires large numbers of lymphocytes with appropriate homing and effector functions directed against malignant cells. Recent advances have largely centered on malignant melanoma, but as the science behind this powerful approach is being honed, the possibility of widening its scope to encompass a broad range of tumor types is becoming a reality. This chapter examines the successful clinical application of ACT in malignant melanoma through the use of tumor-infiltrating lymphocytes (TILs) and how this knowledge is being applied to other tumor types, focusing particularly on the use of T-cell receptor (TCR) and chimeric antigen receptor (CAR)-engineered T cells. It also explores the significant difficulties associated with rolling-out this complex technology, looking particularly at the related toxicities and how future trials may be designed to maximize any benefits whilst limiting these toxicities.

6.2 Background

ACT as a concept was developed during the 1960s. It was subsequently shown that growth of established transplants of chemically induced sarcomas could be inhibited in syngeneic mice by intravenous injection of lymph-node cells from immunized donors.[1] The application of this approach was, however, limited by the need for immunization of lymphocyte donors.

In the allogeneic setting, the role of ACT is well established through the use of donor lymphocyte infusions (DLIs), for example in patients with relapsed leukemia following bone-marrow transplantation. It is thought that lymphocytes from the allogeneic donor respond to either major or minor histocompatibility complex mismatches when infused into the recipient. This response results in a graft-versus-leukemic (GVL) effect with subsequent elimination of the malignant leukemic cells. However, by the same mechanism, the DLI can also respond to normal host tissue resulting in graft-versus-host disease (GVHD) with its associated high level of morbidity and mortality. T cells have been shown to be the critical component within the DLI since their depletion results in loss of both the GVL effect and GVHD. In recent years, a number of strategies have been developed aimed at separating GVL and GVHD. These include reduced intensity conditioning and improved T-cell depletion of the initial donor stem cells followed by DLI.

The powerful anti-tumor effect of the alloresponse has also been explored in clinical studies examining the role of allogencic transplantation in patients with solid tumors. Promising results were noted particularly in patients with renal cell carcinoma (RCC)[2]; however, clinical remissions were only achieved in patients who developed GVHD. This has led to the suggestion that using allogeneic hematopoietic cell transplantation to eliminate tumor without causing GVHD may be restricted to hematological malignancies, such as leukemias and lymphomas, where the alloresponse can be confined to the lymphohematopoietic system.[3] There has therefore been significant interest in the use of autologous cells in ACT for solid malignancies.

Early strategies in the autologous setting concentrated on non-antigen-specific cytotoxicity, for example the use of lymphokine-activated killer (LAK) cells. In animal models, LAK cells can kill tumor cells in a major histocompatibility complex (MHC)-unrestricted manner in response to interleukin-2 (IL-2). They display a wide spectrum of cytotoxicity and initial clinical trials appeared to show promising results when LAK cells were adoptively transferred in conjunction with high-dose infusional IL-2. Unfortunately, a randomized trial of IL-2 plus LAK infusion versus IL-2 alone showed neither a significantly higher response rate (27% vs 18%; two-sided $p = 0.16$) nor a significant improvement in survival (36-month survival 31% vs 17% $p = 0.089$).[4] Trends toward any benefits were felt to be at the expense of severe toxicity and subsequently the field moved toward targeting specific tumor antigens and defining effector-cell populations more closely.

It is well recognized that a variety of immunosuppressive factors exist within the tumor microenvironment that enable malignant cells to flourish whilst suppressing or evading tumor immune surveillance. This process involves both "immunologic sculpting"[5] of the tumor, such as loss of HLA expression, and changes within the patient's immune system, leading to a dominance of regulatory overactivating mechanisms. For example, many cancer patients show an increased proportion of natural regulatory T lymphocytes (Treg) in both the tumor microenvironment and the peripheral blood, which correlates with survival.[6,7] An important factor in the success of TIL therapy described below has been the combination of both adoptive transfer of tumor-reactive lymphocytes in parallel with conditioning chemo-therapy to target these regulatory forces. By depleting Treg and other lymphocytes, conditioning chemotherapy also reduces competition for homeostatic cytokines such as interleukin-7 (IL-7) and interleukin-15 (IL-15). Thus, adoptively transferred TILs have improved access to these cytokines with resulting benefits in terms of *in vivo* expansion and engraftment, key components in the success of this therapy.

6.3 TIL Therapy

Over 20 years ago, it was demonstrated that lymphocytes infiltrating metastatic melanoma deposits (TIL) could be grown in media contain-ing IL-2 and exhibited MHC-restricted recognition of both fresh and cultured melanoma cells.[8] Early clinical studies involved patients with metastatic melanoma receiving cultured TIL either without a prepara-tive chemotherapy regimen or with preinfusion low dose cyclophos-phamide (25 mg/kg), followed in all cases by intravenous IL-2 given at 720,000 IU/kg every 8 h to tolerance.[9] Whilst clinical responses were observed, these tended to be short-lived and at 1 week following infusion, less than 0.1% of cells in the circulation were transferred cells. However, significantly, these early trials demonstrated that objective responses correlated with shorter cell culture time and shorter doubling times of the TIL ($p = 0.0001$ and 0.03 respectively).

It has been suggested from animal models that ACT could be improved by administering lymphodepleting chemotherapy before cell therapy.[10] In a further TIL study from Rosenberg's group at the NIH, this possibility was explored by administering escalating doses of a non-myeloablative chemotherapy regimen consisting of up to 60 mg/kg cyclophosphamide given on 2 consecutive days followed by 5 days of 25 mg/m^2 fludarabine.[11] Cloned melanoma-reactive TIL cultures were administered the next day and followed once again by intravenous IL-2 720,000 IU/kg every 8 h to tolerance. Disappointingly, no objective responses were found and within 2 weeks, the transferred TILs could not be detected in the peripheral blood. It was hypothesized that the absence of CD4$^+$ helper cells within the transferred CD8$^+$ cloned populations may have limited their *in vivo* activation and persistence. The protocol was therefore modified to administer mixed populations of both CD4$^+$ and CD8$^+$ TILs that were selected for being highly reactive against melanoma antigens, but were not cloned[12] (Figure 6.1). This also had the advantage of reducing the intensity of the *ex vivo* stimulation and expansion process resulting in the transfer of cells with an improved activity and proliferation potential. In the first published series with the new protocol, 13 HLA A2$^+$ patients with metastatic melanoma were treated, all of whom had disease refractory to standard therapies including high-dose IL-2.[12] The patients received an average of 7.8×10^{10} cells (range 2.3×10^{10} to 13.7×10^{10}). There were six objective responses with cancer regression at a variety of sites including the lungs, liver, intraperitoneal masses, lymph nodes, and skin. There were associated autoimmune effects (*vitiligo* or uveitis) in five patients, all of whom had evidence of clinical response. A number of patients with cancer regression had durable benefit (12 months or more) and importantly clinical responses strongly correlated with persistence of the transferred cells. The impressive response rate of around

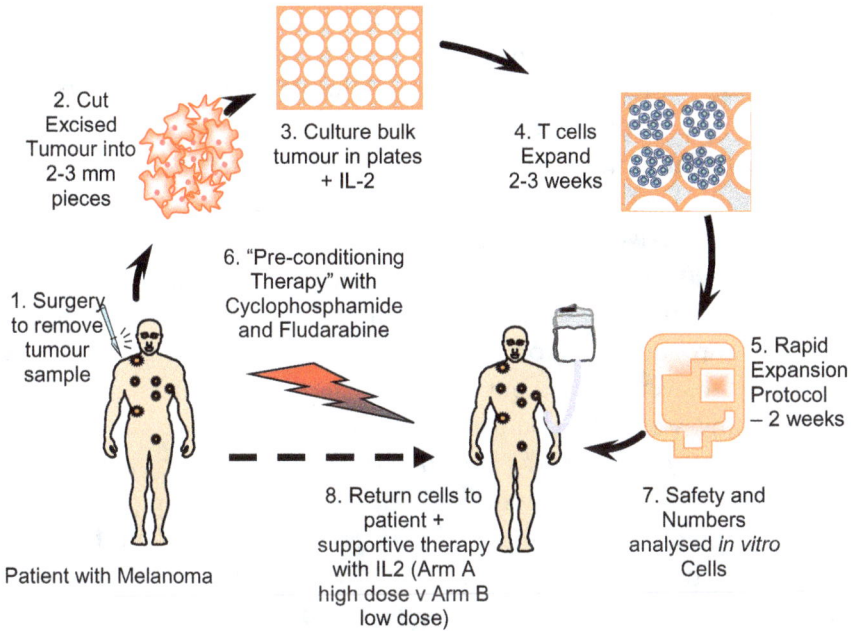

Figure 6.1. Summary Diagram of Treatment Process and Related Procedure

50% was subsequently maintained when the series was expanded to 35 patients[13] and beyond. Later analysis has also shown a strong correlation between the likelihood of response and markers of "younger cells", including telomere length and high surface expression of CD27.[14,15]

A subsequent study attempted to improve *in vivo* cell persistence by transferring TILs that were genetically engineered to produce IL-2.[16] Only two out of five patients in the cohort which also received exogenous IL-2 had objective responses and it was felt by the authors that this was not substantially different from the response rate seen for untransduced TILs. A number of factors were suggested as to why this might have been the case, but an important aspect was felt to be the prolonged culture time that was required to generate sufficient cells for treatment resulting in shortened telomeres.

Animal models have suggested that increasing the intensity of the conditioning regime may further improve the effectiveness of ACT.[17]

Recent trials from Rosenberg's group have therefore escalated the conditioning regime and incorporated the administration of CD34[+] stem hematopoietic stem cells due to the risk of myeloablation.[18] A total of 25 patients received the same chemotherapy as previously described (but condensed to a 5-day period) followed by 2 Gy total body irradiation (TBI) the day before TIL administration. In this case, the objective response rate was 52%; however, when the TBI was increased to 12 Gy in a further 25 patients, the response rate was 72%. This is not statistically significant to the response rate for the 2 Gy group and it is worth noting that the two TBI dose groups (2 or 12 Gy) were carried out as separate non-randomized pilot trials, making any statistical comparison difficult to interpret due to the potential for selection bias. The authors did, however, comment that multiple features of the infused TIL cultures were compared both between the different groups and with patient response, but the only treatment characteristic that was significantly correlated with clinical outcome was the number of IL-2 doses tolerated by patients. The responding patients (in either TBI group as well as those who received conditioning chemotherapy alone) tolerated significantly less IL-2 than the non-responding patients. It was suggested that this may be as a result of the TILs being effectively engaged in anti-tumor responses with resulting release of secondary cytokines into the host's circulation.[18] Any improvement in response rate has not yet translated into a statistically significant benefit in survival amongst patients who received no TBI versus 2 Gy versus 12 Gy ($p = 0.13$), although with a median follow-up of only 10 months in the 12 Gy treatment group, further follow-up is clearly required before any firm conclusion can be drawn.

A major issue that will determine whether or not this type of TIL ACT will, in the future, become widely available to patients around the world is whether the significant associated toxicity can be widely dealt with. In the trials described above, the 25 patients who received 12 Gy TBI had increased toxicity compared to the earlier patients who had non-myeloablative conditioning chemotherapy.[18] These toxicities included fatigue, anorexia, and weight loss with four patients requiring intubation and ventilation for somnolence. There was

Table 6.1. Limitations of TIL ACT.

Limitation	Description
Restricted to melanoma	Only currently applicable to melanoma, although other tumor types such as RCC have been shown to contain TIL.[25] Tumor-specific T cells derived from peripheral blood is one strategy that might open up ACT to other tumor types. The use of autologous Melan-A-specific cytotoxic T-lymphocyte clones derived from peripheral blood was recently explored by Khammari and colleagues.[56]
HLA A2 restriction	Restricted to HLA A2 positive patients.
Failure to expand TILs	TILs can only be expanded in approximately 50% of cases.
Cost	Requires significant investment in infrastructure, equipment, and consumables as well as staffing costs.
Expertise	Requires considerable laboratory and clinical expertise.
Lack of commercial interest	Pharmaceutical and biotechnology companies seek "off-the-shelf" drugs that are easy to manufacture and administer.
Regulation	Complex and expensive.

significant bone marrow toxicity with many patients requiring both platelet and red blood cell support and 16 out of 25 patients suffering common toxicity criteria adverse event (CTCAE) grade 3 or 4 febrile neutropenia. It was reported that patients generally returned to normal daily routines by 2 to 3 months after treatment.

The data emerging from Rosenberg's group at the NIH appears to show convincing evidence that TIL therapy can induce durable objective responses in patients with metastatic melanoma and, along with preclinical data, provide the rational for the use of preconditioning chemotherapy. However, given the severe toxicities, there is a pressing need for fully randomized clinical trials to provide conclusive evidence of benefit to this group of patients and to explore whether there is any supplemental benefit from the addition of TBI to the conditioning regime. In addition, although the potency of ACT is becoming increasingly apparent, the strategy remains technically challenging; TILs have to be isolated from surgical tumor specimens, which is particularly difficult in the case of solid tumors, and tumor-reactive T cells have to be identified and expanded to sufficient numbers for therapeutic application.

The engraftment benefits of using "young" T cells noted in earlier trials have also been seen in more recent trials. This has led to trials specifically examining the use of "young TILs" using cells that have not been selected for specificity at all. Impressively, this maintains response rate at around 50%, although duration of response remains to be seen either with all cells[19-21] or with CD8-selected TIL (Dudley — personal communication). This more straightforward approach clearly warrants further assessment although a key unanswered question is which T-cell reactivities within the TIL are involved in the observed clinical responses. If these can be defined, for example through the use of high-throughput T-cell immunomonitoring,[22] it may be possible to produce more defined T-cell products that are enriched for these specificities.

6.4 Engineered T Cells — Clinical Trials

TIL therapy is currently limited to a small minority of all patients with cancer, essentially a subset of melanoma patients who have preexisting tumor-reactive lymphocytes that can be expanded *ex vivo*. However, it appears that virtually all tumor types are susceptible to cytokine release and lysis by lymphocytes upon recognition of the appropriate tumor-associated antigen (TAA). One approach that is being actively pursued to broaden the applicability of ACT is the introduction of genes into peripheral blood-derived lymphocytes to provide the cells with reactivity against a specified TAA. Two strategies show particular promise: (i) the use of cloned TCR and (ii) chimeric antibody receptors (CARs) as discussed in Chapters 3 and 4, respectively. Discussed in this chapter are some of the clinical strategies being adopted for these approaches, what toxicities have been observed, and how they may be improved in the future (summarized in Table 6.2).

Table 6.2. Potential improvements in ACT.

Improvement	Potential strategies
Increasing diversity of target tumors	TILs from tumors other than melanoma. Use of T lymphocytes genetically modified with CAR or TCR.
Improving efficacy of transferred cells	Use of "young" T cells. Optimization of *ex vivo* cell culture to manipulate final T-cell phenotype toward improved *in vivo* activity. Use of CAR/TCR with appropriate antigen avidity. Use of second- or third-generation CAR incorporating additional co-stimulatory molecules, cytokines, homing molecules, and/or anti-apoptotic molecules. Co-stimulation through virus (e.g., EBV) specific T cells. Block inhibitory signals from regulatory T cells, for example the use of antibodies to CTLA4 or PD1.
Minimization of toxicity	Selection of appropriate TAA to minimize cross-reactivity resulting from expression on normal tissues. TCR engineering to prevent mis-pairing of α or β chains. Careful trial design with appropriate starting dose of T cells and dose escalation. Strategies to eliminate adoptively transferred cells should toxicity be experienced (e.g., incorporation of suicide genes). Exploration of regimes with reduced toxicity (and/or improved efficacy) conditioning chemotherapy and cytokine administration.
Maximizing progress	Trial design incorporating standard assays and reporting. Trials designed within consortia which limit the number of variables to improve cross-trial comparisons.

6.4.1 *Clinical trials of TCR-engineered T cells*

The first clinical trial to successfully show objective responses to ACT using genetically engineered lymphocytes was published by Morgan and colleagues in 2006.[23] A total of 17 patients with progressive metastatic melanoma were given the same non-myeloablative preconditioning chemotherapy (fludarabine plus cyclophosphamide) along with high-dose IL-2 post-cell infusion as used in the successful TIL trials described above. Autologous peripheral blood lymphocytes which had been transduced with genes encoding the alpha and beta chains of an anti-MART-1 TCR were. Gene transfer efficiencies were between 17 and 67% (mean 42%). An initial cohort of patients received cells that had been cultured *ex vivo* for an extended period of 19 days. Poor long-term persistence of transduced cells was noted ($\leq 2\%$ beyond 50 days) and there were no clinical responses in this group. However, in a later cohort where the cells were cultured for a shorter period (about 7 days), all eight patients who provided samples at >50 days after treatment exhibited durable transduced cell persistence of >17%. Two patients, both of whom had rapidly progressive disease when they were treated, experienced significant and durable clinical responses. Both also had sustained high levels of transduced cells at 1 year after infusion (20% and 70% of total body lymphocytes). This trial not only highlights the potential use of genetically modified lymphocytes as a cancer therapy, but also reiterates the importance of short *ex vivo* culture and sustained persistence of infused lymphocytes as key factors for success of this type of therapy.

Other trials targeting melanoma antigens have since been performed including high avidity MART-1 TCR and a TCR directed against the antigen gp-100. Of note, TCRs with high affinity have been seen to cause pathology consistent with "on-target" toxicity (for example inner ear and retinal toxicity) due to antigen expression in these organs at similar levels as on targeted melanoma cells.[24] While manageable in this particular case, the observation suggests that great care must be taken in the selection of target antigens. In addition to the potential for "on-target" toxicity, the concern has been raised that toxicity may occur through various "off-target" mechanisms and these are discussed in Sec. 6. A number of TCRs that recognize TAAs expressed on a broad range of tumor types are also being explored in planned or ongoing clinical trials. These TAAs include NY-ESO-1, CEA, p53, and HER2.[25]

6.4.2 Clinical trials of CAR-Engineered T cells

As previously discussed in Chapter, it is possible to modify primary human lymphocytes from a cancer patient with a recombinant chimeric immune receptor (CIR). The CIR is composed of an extracellular, antibody-derived antigen-binding domain fused to an intracellular TCR-derived CD3ζ domain to provide T-cell activation upon engagement of antigen. By utilizing an antibody for binding, CIR recognition is independent of processes frequently modified in malignant cells such as antigen processing and presentation in the MHC. In addition, non-classical T-cell antigens, such as carbohydrates, can be recognized.

Early clinical trials of CAR-engineered T cells showed little evidence for efficacy or toxicity. In these trials, cells survived *in vivo* for short periods of time, most likely due to suboptimal *ex vivo* growth conditions combined with a lack of preconditioning chemotherapy.[26–28] However, under these conditions, it is apparent that even short-term persistence can produce significant on-target toxicity in cases where tumor-selective antigen is also expressed on cells that are accessible to infused cells.[27] This is further discussed in Sec. 6.

First-generation CARs[29,30] which consist of only an extracellular antibody-derived antigen-binding domain fused to an intracellular TCR-derived CD3ζ have been shown to produce only suboptimal cellular activation of cells as co-stimulation is not provided. Second-generation CARs consequently possess a co-stimulatory moiety, for example derived from CD28, OX40 (CD134), or 4-1BB (CD137) together with the CD3ζ intracellular domain to improve T-cell activation and expansion. In order to improve this even further, there has been the development of third-generatrion CARs which combine co-stimulatory domains such as CD28 and 4-1BB.[30,31] These receptors have greater efficacy in preclinical testing[32–34] and an increasing number of clinical trials are currently evaluating second- and third-generation CARs. However, the potential for severe toxicity with these increasingly powerful receptors is clear and current preclinical models are not necessarily adequate to assess their full impact. Unfortunately as discussed in Sec. 6.5, there have been recent reports of deaths in such clinical trials.

6.5 Toxicities and Safety of ACT

Many TAAs are expressed at low levels in normal tissues. This brings with it the potential for cross-reactivity of any adoptively transferred T cells which have specificity for a TAA (on-target toxicity). For example, in the TIL trials, both anterior uveitis and *vitiligo* have been noted as a result of cross-reactivity to normal MART-1 antigen expressed at these sites.[12] The uveitis was controlled with steroid eye drops and the *vitiligo* did not require treatment.

Nevertheless, cross-reactivity may not always be so indolent. In an ACT trial of CAR-transduced T cells directed against an epitope (G250) on carboxy-anhydrase-IX (CAIX), all three RCC patients developed liver enzyme disturbances reaching NCI CTCAE grades 2 to 4.[27] This occurred despite a lack of evidence for significant cell persistence or clinical efficacy in this trial. One patient required treatment with corticosteroids, whilst the remaining two patients' abnormal blood tests settled on discontinuation of the T-cell infusions. The liver enzyme disturbances were, however, sufficient to result in the trial being discontinued. It is known from labeling experiments that infused cells localize non-specifically in lung and liver immediately after infusion.[35] Expression of G250 has been noted in bile duct epithelial cells and on-target cross-reactivity of the transferred cells with this expressed antigen was thought to be the most likely cause for the observed cholangitis and damage to the bile duct epithelium. This observation illustrates not only the potential power of the approach, but also the need to consider carefully the choice of target antigen, including the site and accessibility of any normal antigen expression as well as its extent.

A relevant example of an antigen that is actively being targeted in both TCR and CAR ACT strategies is that of carcinoembryonic antigen (CEA).[36] CEA is highly expressed in a number of tumor types, particularly those in the gastrointestinal tract such as colorectal adenocarcinoma. It is also expressed at low levels in some normal tissues such as the large bowel[37] where expression occurs on the luminal aspect of the cells, which may afford some protection from the gene-modified cells, but the possibility of bowel autoimmunity in these trials cannot be excluded.

As the technology surrounding gene-modified T cells becomes more powerful, the potential threat of on-target toxicity increases since no tumor antigens (except viral antigens or unique mutations) are truly tumor specific. This is illustrated with the development of second- and third- generation CARs which encode multiple signaling domains, including co-stimulatory molecules[32] (e.g., CD28 and 4-1BB) and also receptors that incorporate cytokines[16] (e.g., IL2, IL7 and IL15), homing molecules (e.g., CCR7), and molecules to prevent apoptosis[38] (e.g., BCL2). A recent preclinical report highlighted the potential for both short- and long-term toxicity mediated by CAR-modified T cells.[39] In this fully autologous mouse model, murine CD19 was targeted with murine T cells transformed with a second-generation CAR incorporating CD28 and CD3ζ. This was more effective in controlling target cells, but also produced increased short-term toxicity in the form of cytokine release syndrome and in the longer-term, continued stimulation of engineered T cells resulted in uncontrolled proliferation with gross-lymphocyte accumulation.

Unfortunately, the risk of such on-target toxicity with these second- and third-generation CARs is becoming apparent in clinical trials with two recent reports of the deaths of patients.[40,41] In the case reported by Brentjens and colleagues, a second-generation CAR targeting CD19 and incorporating CD28 was administered. While patients in the first planned cohort did not have serious adverse events, the first patient to receive cyclophosphamide preconditioning chemotherapy developed significant acute problems. The patient had a diagnosis of chronic lymphocytic leukemia (CLL) and had previously been heavily pretreated with chemotherapy. They received an infusion of 3×10^7 T cells per kg and subsequently developed hypotension, dyspnoea, and fever and unfortunately died shortly after. A post-mortem did not reveal the precise cause and whilst the investigators proposed that low-grade sepsis may have triggered the observed elevated cytokine levels before T-cell infusion, the concern exists that the transferred modified T cells may have contributed. In the second case report from Morgan and colleagues at the Surgery Branch (NIH), a patient with refractory colon cancer with lung and liver metastases received a third-generation CAR targeting HER2/neu/Erb B2 with both CD28 and 4-1BB co-stimulation. The patient received 10^{10} T cells following non-myeloablative lymphodepletion and immediately developed significant pulmonary toxicity and a cytokine storm followed by cardiac arrest and death 5 days after treatment. Detailed analysis showed increased levels of IFN-γ, GM-CSF, TNF-α, IL-6, and IL-10 and pulmonary infiltrates. The overall effect was thought to have resulted from low-level expression of Her2 in the normal lung where the cells localize after infusion. In both cases, the investigators are reducing the numbers of T cells given before continuing; however, the pharmacodynamics of engineered T cells are clearly much more difficult to predict than of classical drugs or antibodies.

Autoimmunity could also theoretically arise in ACT strategies from the chance expansion of T cells that have natural reactivity to a self-antigen. Such cells might normally be tolerized, but if transduced with a TCR or CAR to a TAA, they may be activated on recognition of the TAA and subsequently display cytotoxicity toward cells expressing the original self-antigen. The most likely organs to be affected may be those most frequently associated with autoimmune attack (e.g., kidney, liver, thyroid, joints, skin, and lungs).

A further potential source of toxicity is specific to engineered TCR-based strategies; that of introduced TCR α or β chains mispairing with endogenous α or β TCR chains. This could result in either reduced reactivity to the targeted TAA, or potentially more seriously, reactivity to an unknown, unintended alternative antigen. However, a number of approaches to prevent mis-pairing such as genetic modification of the TCR constant regions are being adopted and these are discussed in detail in Chapter 4.

As already mentioned above, the most successful ACT TIL trials to date emerging from the NIH were associated with significant toxicity from the conditioning chemotherapy and TBI regimes and the supplemental high-dose IL-2. The group at the NIH clearly have extensive experience of managing these patients, but it remains to be seen whether the considerable toxicities associated with such treatments can be appropriately managed in smaller (albeit specialized) units. Whilst preclinical data suggests that conditioning chemotherapy is beneficial to transferred cell engraftment,[10] it may be possible to reduce the dose of IL-2, for example to a more manageable regime of 125,000 IU/kg subcutaneously.[25] This would significantly reduce the toxicity compared to high-dose intravenous IL-2 and could potentially be administered on an outpatient basis.

A high-profile potential risk of ACT using gene-modified T cells is that of oncogenesis resulting from insertional mutagenesis as part of the retroviral gene transfer process. Insertional mutagenesis has been documented in the case of engineered stem cell therapy for the inherited immune deficiency states X-linked SCID[42] and chronic granulomatous disease.[43] In contrast, gene transfer into differentiated T lymphocytes has a good safety record from studies over the last two decades. The first gene marking study in T cells was reported in 1990 using retroviruses to mark TILs for the treatment of melanoma[44] and since then a number of reports have demonstrated long-lived gene-modified T-cell engraftment without evidence of resulting oncogenesis.[45-47] However, close monitoring of patients who receive gene-modified T cells for clonal expansion of T cells is clearly an important aspect of safety monitoring for any such trial. Ongoing work is directed toward reducing the theoretical risk of insertional mutagenesis.

6.6 Future Strategies

6.6.1 *T-cell engineering*

The recent serious adverse events that resulted in two deaths in clinical trials involving gene-modified T cells, both involved CARs that incorporated co-stimulatory molecules to optimize cytokine secretion and cytolysis. In the case of the NIH trial, T-cell infusion lead immediately to a rapid cytokine storm. The power and potential unpredictability of these molecules clearly needs to be carefully balanced with the desire for efficacy in future clinical trials. One alternative way to provide a co-stimulatory signal is by making use of virus-specific T cells engineered to express a first-generation CAR. The engineered T cells then receive co-stimulation by antigen binding through their native TCR on antigen-presenting cells (APCs).[48] Evidence that this could be an effective strategy comes from a recent trial in which T cells that were Epstein–Barr virus (EBV)-specific were genetically modified with a neuroblastoma-associated antigen-specific CAR. These cells persisted longer *in vivo* than unselected T cells with the same CAR which did not receive co-stimulation by EBV antigen presenting APCs.[49]

In circumstances where adoptively transferred cells are causing significant on-target toxicity, it may become necessary to attempt to eradicate them from the circulation. Administration of steroids was felt to be useful in the clinical trial where liver toxicity was noted using T cells expressing a CAR targeting G250.[27] However, several more specific strategies have the potential to be helpful. For some time, the possibility of engineering cells with a herpes simplex thymidine kinase (TK) gene has been investigated. This would in theory allow elimination of the gene-modified T cells by administration of nucleoside prodrugs. However, use of the TK gene alone may be of limited value due to insufficient kinase activity, which can be at least partly compensated by the use of a second suicide gene.[50] Another approach includes the use of inducible caspase-9 expressed in gene-modified T cells to induce apoptosis. Induction occurs when the active enzyme is formed out of two non-functional molecules on administration of a dimerizer.[51] A further strategy involves incorporating a myc-tag into transgenic TCR-α chain which can then be eliminated *in vivo* by administration of a depleting anti-myc antibody.[52] An alternative to introducing molecules to induce cell death may be the use of transient expression of CARs, for example by electropulsing T cells with coding RNA.[53] The RNA-modified T cells gradually loose CAR expression when they are expanding (half-life approximately 2 days) and this strategy may be particularly useful in exploring acute toxicity in the first-in-man setting.

To date, the majority of CARs that have been developed have made use of antibodies with high affinity binding domains. While this is attractive in terms of sensitivity, high affinity binding discriminates less well between moderate/low and high-level antigen expression in physiological cells versus target cells respectively. This could increase the risk of activation upon binding to non-malignant cells expressing moderate/low levels of antigen. Increasing the binding affinity above threshold does not increase T-cell activation in terms of cytokine release and cell kill, but loses selectivity for high-level antigen expressing targets.[54] Future approaches therefore could make use of moderate (or even low) affinity binding CARs which will only trigger T-cell activation when bound to high levels of antigen available on the tumor cells.

6.6.2 *Clinical trial strategies*

If the ultimate goal is to achieve effective therapy funded by healthcare providers, then a number of hurdles must be overcome. Growing cells for ACT carries significant cost in terms of staffing and equipping appropriate GMP facilities. Administration of the cells must be carried out in suitable clinical facilities with access to intensive care when required. Over and above the financial costs, a high level of both laboratory and clinical expertise is required. This goes a long way to explaining why, to date, TIL-based ACT has only been established at a few specialized centers such as the NIH and highlights the need for a co-ordinated approach to developing the technology if it is ever to become widely available. A key factor is the development of focused consortia such as the European-based ATTACK consortium (www.ATTACK-cancer.org). The ATTACK group has made a number of recommendations, including proposing that TIL trials in melanoma should be carried out in a multicenter setting preferably in randomized trials. If these trials confirm the benefits seen in single center trials, the methodology could then be further optimized through larger scale trials and consideration given to exploring the approach in renal and possibly other cancers.

The recent deaths in clinical trials illustrate the importance of careful design of future ACT clinical trials for both natural and engineered T cells. The multitude of potential clinical targets needs to be carefully investigated prior to first-in-man trials in a number of ways, including the development of improved animal models. Other preclinical considerations include assessment of the level, function, and accessibility of antigen expression on normal tissue and the affinity of the receptor for the target. These issues need to be reflected in the trial design with a focus on both the starting dose of T cells and dose escalation. Particular consideration needs to be given in trials using second- or third-generation CARs where the use of suicide genes or other means of abrogating toxicity also need to be considered.

The complex nature of ACT T-cell trials means that at present, there are no standard reporting formats or even accepted assay systems to monitor immune responses. By establishing consortia with shared goals, these hurdles can be overcome in such a way as to facilitate comparison of emerging data across different trials. More fundamentally, however, is the risk that small-scale trials carried out by individual groups would still differ in so many parameters that establishing critical factors in success (or failure) would be severely problematic even in the setting of standard reporting or assays. One proposal to avoid this is that there are agreed formats, for example for receptors, cell growth, and engraftment conditions that are only varied by one or two parameters across trials.[55] The implementation of this proposal is clearly difficult; whilst the strategy is likely to be of great benefit to the community as a whole, the direct gain for any individual is reduced and this would risk compliance. Funding bodies may well have a key role to play if it is ever to become a reality.

6.7 Conclusions

There have been many developments in the treatment of metastatic cancer in recent years; however, most novel agents only provide a modest improvement in survival. In contrast, substantial benefits in terms of durable remissions have been seen in some patients receiving TIL therapy for metastatic melanoma. This makes further development of ACT using natural T cells in melanoma and for a broad range of tumors using gene-modified T cells an attractive prospect in spite of the complex nature and cost of the strategy.

References

1. Borberg, H., Oettgen, H.F., Choudry, K., *et al.* (1972). Inhibition of established transplants of chemically induced sarcomas in syngeneic mice by lymphocytes from immunized donors. *Int J Cancer* 10, 539–547.

2. Childs, R., Chernoff, A., Contentin, N., et al. (2000). Regression of metastatic renal-cell carcinoma after nonmyeloablative allogeneic peripheral-blood stem-cell transplantation. N Engl J Med 343, 750–758.

3. Mapara, M.Y. and Sykes, M. (2004). Tolerance and cancer: Mechanisms of tumor evasion and strategies for breaking tolerance. J Clin Oncol 22, 1136–1151.

4. Rosenberg, S.A., Lotze, M.T., Yang, J.C., et al. (1993). Prospective randomized trial of high-dose interleukin-2 alone or in conjunction with lymphokine-activated killer cells for the treatment of patients with advanced cancer. J Natl Cancer Inst 85, 622–632.

5. Dunn, G.P., Old, L.J., and Schreiber, R.D. (2004). The immunobiology of cancer immunosurveillance and immunoediting. Immunity 21, 137–148.

6. Griffiths, R.W., Elkord, E., Gilham, D.E., et al. (2007). Frequency of regulatory T cells in renal cell carcinoma patients and investigation of correlation with survival. Cancer Immunol Immunother 56, 1743–1753.

7. Sasada, T., Kimura, M., Yoshida, Y., et al. (2003). CD4+CD25+ regulatory T cells in patients with gastrointestinal malignancies: Possible involvement of regulatory T cells in disease progression. Cancer 98, 1089–1099.

8. Muul, L.M., Spiess, P.J., Director, E.P., et al. (1987). Identification of specific cytolytic immune responses against autologous tumor in humans bearing malignant melanoma. J Immunol 138, 989–995.

9. Rosenberg, S.A., Yang, J.C., Topalian, S.L., et al. (1994). Treatment of 283 consecutive patients with metastatic melanoma or renal cell cancer using high-dose bolus interleukin 2. JAMA 271, 907–913.

10. Cheadle, E.J., Gilham, D.E., and Hawkins, R.E. (2008). The combination of cyclophosphamide and human T cells genetically engineered to target CD19 can eradicate established B-cell lymphoma. Br J Haematol 142, 65–68.

11. Dudley, M.E., Wunderlich, J.R., Yang, J.C., et al. (2002). A phase I study of nonmyeloablative chemotherapy and adoptive transfer of autologous tumor antigen-specific T lymphocytes in patients with metastatic melanoma. J Immunother 25, 243–251.

12. Dudley, M.E., Wunderlich, J.R., Robbins, P.F., et al. (2002). Cancer regression and autoimmunity in patients after clonal repopulation with antitumor lymphocytes. Science 298, 850–854.

13. Dudley, M.E., Wunderlich, J.R., Yang, J.C., et al. (2005). Adoptive cell transfer therapy following non-myeloablative but lymphodepleting chemotherapy for the treatment of patients with refractory metastatic melanoma. J Clin Oncol 23, 2346–2357.

14. Zhou, J., Shen, X., Huang, J., et al. (2005). Telomere length of transferred lymphocytes correlates with in vivo persistence and tumor regression in melanoma patients receiving cell transfer therapy. J Immunol 175, 7046–7052.

15. Huang, J., Kerstann, K.W., Ahmadzadeh, M., et al. (2006). Modulation by IL-2 of CD70 and CD27 expression on CD8+ T cells: Importance for the therapeutic effectiveness of cell transfer immunotherapy. J Immunol 176, 7726–7735.

16. Heemskerk, B., Liu, K., Dudley, M.E., *et al.* (2008). Adoptive cell therapy for patients with melanoma, using tumor-infiltrating lymphocytes genetically engineered to secrete interleukin-2. *Hum Gene Ther* 19, 496–510.

17. Wrzesinski, C., Paulos, C.M., Kaiser, A., *et al.* Increased intensity lymphodepletion enhances tumor treatment efficacy of adoptively transferred tumor-specific T cells. *J Immunother* 33, 1–7.

18. Dudley, M.E., Yang, J.C., Sherry, R., *et al.* (2008). Adoptive cell therapy for patients with metastatic melanoma: Evaluation of intensive myeloablative chemoradiation preparative regimens. *J Clin Oncol* 26, 5233–5239.

19. Besser, M.J., Shapira-Frommer, R., Treves, A.J. *et al.* (2009). Minimally cultured or selected autologous tumor-infiltrating lymphocytes after a lymphodepleting chemotherapy regimen in metastatic melanoma patients. *J Immunother* 32, 415–423.

20. Markel, G., Cohen-Sinai, T., Besser, M.J., *et al.* (2009). Preclinical evaluation of adoptive cell therapy for patients with metastatic renal cell carcinoma. *Anticancer Res* 29, 145–154.

21. Besser, M.J., Shapira-Frommer, R., Treves, A.J., *et al.* (2010). Clinical responses in a phase II study using adoptive transfer of short-term cultured tumor infiltration lymphocytes in metastatic melanoma patients. *Clin Cancer Res* 16, 2646–2655.

22. Hadrup, S.R., Bakker, A.H., Shu, C.J., *et al.* (2009). Parallel detection of antigen-specific T-cell responses by multidimensional encoding of MHC multimers. *Nat Methods* 6, 520–526.

23. Morgan, R.A., Dudley, M.E., Wunderlich, J.R. *et al.* (2006). Cancer regression in patients after transfer of genetically engineered lymphocytes. *Science* 314, 126–129.

24. Johnson, L.A., Morgan, R.A., Dudley, M.E., *et al.* (2009). Gene therapy with human and mouse T-cell receptors mediates cancer regression and targets normal tissues expressing cognate antigen. *Blood* 114, 535–546.

25. Rosenberg, S.A. and Dudley, M.E. (2009). Adoptive cell therapy for the treatment of patients with metastatic melanoma. *Curr Opin Immunol* 21, 233–240.

26. Kershaw, M.H., Westwood, J.A., Parker, L.L. *et al.* (2006). A phase I study on adoptive immunotherapy using gene-modified T cells for ovarian cancer. *Clin Cancer Res* 12, 6106–6115.

27. Lamers, C.H., Sleijfer, S., Vulto, A.G. *et al.* (2006). Treatment of metastatic renal cell carcinoma with autologous T-lymphocytes genetically retargeted against carbonic anhydrase IX: first clinical experience. *J Clin Oncol* 24, e20–e22.

28. Lamers, C.H., van Elzakker, P., Langeveld, S.C. *et al.* (2006). Process validation and clinical evaluation of a protocol to generate gene-modified T lymphocytes for imunogene therapy for metastatic renal cell carcinoma: GMP-controlled transduction and expansion of patient's T lymphocytes using a carboxy anhydrase IX-specific scFv transgene. *Cytotherapy* 8, 542–553.

29. Gross, G., Waks, T., and Eshhar, Z. (1989). Expression of immunoglobulin-T-cell receptor chimeric molecules as functional receptors with antibody-type specificity. *Proc Natl Acad Sci USA* 86, 10024–10028.

30. Eshhar, Z., Waks, T., Gross, G. *et al.* (1993). Specific activation and targeting of cytotoxic lymphocytes through chimeric single chains consisting of antibody-binding domains and the gamma or zeta subunits of the immunoglobulin and T-cell receptors. *Proc Natl Acad Sci USA* 90, 720–724.

31. Zhao, Y., Wang, Q.J., Yang, S. *et al.* (2009). A herceptin-based chimeric antigen receptor with modified signaling domains leads to enhanced survival of transduced T lymphocytes and antitumor activity. *J Immunol* 183, 5563–5574.

32. Haynes, N.M., Trapani, J.A., Teng, M.W. *et al.* (2002). Single-chain antigen recognition receptors that costimulate potent rejection of established experimental tumors. *Blood* 100, 3155–3163.

33. Brentjens, R.J., Latouche, J.B., Santos, E. *et al.* (2003). Eradication of systemic B-cell tumors by genetically targeted human T lymphocytes co-stimulated by CD80 and interleukin-15. *Nat Med* 9, 279–286.

34. Moeller, M., Haynes, N.M., Trapani, J.A. *et al.* (2004). A functional role for CD28 costimulation in tumor recognition by single-chain receptor-modified T cells. *Cancer Gene Ther* 11, 371–379.

35. Fisher, B., Packard, B.S., Read, E.J. *et al.* (1989). Tumor localization of adoptively transferred indium-111 labeled tumor infiltrating lymphocytes in patients with metastatic melanoma. *J Clin Oncol* 7, 250–261.

36. Sheen, A.J., Irlam, J., Kirillova, N. *et al.* (2003). Gene therapy of patient-derived T lymphocytes to target and eradicate colorectal hepatic metastases. *Dis Colon Rectum* 46, 793–804.

37. Hammarstrom, S. (1999). The carcinoembryonic antigen (CEA) family: Structures, suggested functions and expression in normal and malignant tissues. *Semin Cancer Biol* 9, 67–81.

38. Eaton, D., Gilham, D.E., O'Neill, A. *et al.* (2002). Retroviral transduction of human peripheral blood lymphocytes with Bcl-X(L) promotes *in vitro* lymphocyte survival in pro-apoptotic conditions. *Gene Ther* 9, 527–535.

39. Cheadle, E. (2010). A CD28 containing CAR+ T-cells targeting mouse CD19 causes both short and long-term toxicity when transferred into pre-conditioned mice. *Hum Gene Ther* 20, 1383.

40. Morgan, R.A., Yang, J.C., Kitano, M. *et al.* (2010). Case report of a serious adverse event following the administration of T cells transduced with a chimeric antigen receptor recognizing ERBB2. *Mol Ther* 18, 843–851.

41. Brentjens, R., Yeh, R., Bernal, Y. *et al.* (2010). Treatment of chronic lymphocytic leukemia with genetically targeted autologous T cells: Case report of an unforeseen adverse event in a phase I clinical trial. *Mol Ther* 18, 666–668.

42. Hacein-Bey-Abina, S., Von Kalle, C., Schmidt, M. *et al.* (2003). LMO2-associated clonal T cell proliferation in two patients after gene therapy for SCID-X1. *Science* 302, 415–419.

43. Stein, S., Ott, M.G., Schultze-Strasser, S. *et al.* (2010). Genomic instability and myelodysplasia with monosomy 7 consequent to EVI1 activation after gene therapy for chronic granulomatous disease. *Nat Med* 16, 198–204.

44. Rosenberg, S.A., Aebersold, P., Cornetta, K. *et al.* (1990). Gene transfer into humans-immunotherapy of patients with advanced melanoma, using tumor-infiltrating lymphocytes modified by retroviral gene transduction. *N Engl J Med* 323, 570–578.

45. Blaese, R.M., Culver, K.W., Miller, A.D. *et al.* (1995). T lymphocyte-directed gene therapy for ADA- SCID: Initial trial results after 4 years. *Science* 270, 475–480.

46. Onodera, M., Ariga, T., Kawamura, N. *et al.* (1998). Successful peripheral T-lymphocyte-directed gene transfer for a patient with severe combined immune deficiency caused by adenosine deaminase deficiency. *Blood* 91, 30–36.

47. Muul, L.M., Tuschong, L.M., Soenen, S.L. *et al.* (2003). Persistence and expression of the adenosine deaminase gene for 12 years and immune reaction to gene transfer components: Long-term results of the first clinical gene therapy trial. *Blood* 101, 2563–2569.

48. Cooper, L.J., Al-Kadhimi, Z., Serrano, L.M. *et al.* (2005). Enhanced antilymphoma efficacy of CD19-redirected influenza MP1-specific CTLs by cotransfer of T cells modified to present influenza MP1. *Blood* 105, 1622–1631.

49. Pule, M.A., Savoldo, B., Myers, G.D. *et al.* (2008). Virus-specific T cells engineered to coexpress tumor-specific receptors: Persistence and antitumor activity in individuals with neuroblastoma. *Nat Med* 14, 1264–1270.

50. Uckert, W., Kammertons, T., Haack, K. *et al.* (1998). Double suicide gene (cytosine deaminase and herpes simplex virus thymidine kinase) but not single gene transfer allows reliable elimination of tumor cells *in vivo*. *Hum Gene Ther* 9, 855–865.

51. Hoyos, V., Savoldo, B., Quintarelli, C. *et al.* (2010). Engineering CD19-specific T lymphocytes with interleukin-15 and a suicide gene to enhance their anti-lymphoma/leukemia effects and safety. *Leukemia* 24, 1160–1170.

52. Kieback, E., Charo, J., Sommermeyer, D. *et al.* (2008). A safeguard eliminates T cell receptor gene-modified autoreactive T cells after adoptive transfer. *Proc Natl Acad Sci USA* 105, 623–628.

53. Birkholz, K., Hombach, A., Krug, C. *et al.* (2009). Transfer of mRNA encoding recombinant immunoreceptors reprograms CD4+ and CD8+ T cells for use in the adoptive immunotherapy of cancer. *Gene Ther* 16, 596–604.

54. Chmielewski, M., Hombach, A., Heuser, C. *et al.* (2004). T cell activation by antibody-like immunoreceptors: Increase in affinity of the single-chain fragment

domain above threshold does not increase T cell activation against antigen-positive target cells but decreases selectivity. *J Immunol* 173, 7647–7653.

55. Hawkins, R.E., Gilham, D.E., Debets, R. *et al.* (2010). Development of adoptive cell therapy for cancer: A clinical perspective. *Hum Gene Ther* 21, 665–672.

56. Khammari, A., Labarriere, N., Vignard, V. *et al.* (2009). Treatment of metastatic melanoma with autologous Melan-A/MART-1-specific cytotoxic T lymphocyte clones. *J Invest Dermatol* 129, 2835–2842.

www.ingramcontent.com/pod-product-compliance
Lightning Source LLC
Chambersburg PA
CBHW050602190326
41458CB00007B/2147